Brickwork for Apprentices

Fifth edition

J. C. Hodge L.M.G.B.
Former Senior Lecturer, Metropolitan University, London

For the fifth edition revised by

M. Thorpe BA. L.M.G.B.
Former Head of Construction, West Nottinghamshire College

Previously revised for the fourth edition by R. J. Baldwin,
Former Senior Lecturer, Willesden College of Technology, London

ELSEVIER

Amsterdam • Boston • Heidelberg • London • New York • Oxford
Paris • San Diego • San Francisco • Singapore • Sydney • Tokyo
Butterworth-Heinemann is an imprint of Elsevier

Butterworth-Heinemann is an imprint of Elsevier
Linacre House, Jordan Hill, Oxford OX2 8DP, UK
30 Corporate Drive, Suite 400, Burlington, MA 01803, USA

First edition 1944
Second edition 1960
Third edition 1971
Fourth edition 1993
Reprinted 2001, 2002 (twice), 2003, 2004
Transferred to digital printing 2004, 2005
Fifth edition 2006
Reprinted 2007

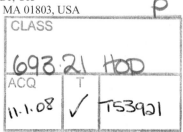
Notice
No responsibility is assumed by the publisher for any injury and/or damage to persons
or property as a matter of products liability, negligence or otherwise, or from any use
or operation of any methods, products, instructions or ideas contained in the material
herein. Because of rapid advances in the medical sciences, in particular, independent
verification of diagnoses and drug dosages should be made

British Library Cataloguing in Publication Data
A catalogue record for this book is available from the British Library

Library of Congress Cataloging-in-Publication Data
A catalog record for this book is available from the Library of Congress

ISBN–13: 978-0-7506-6752-4
ISBN–10: 0-7506-6752-4

For information on all Butterworth-Heinemann publications
visit our website at books.elsevier.com

Printed and bound in *Great Britain*

07 08 09 10   10 9 8 7 6 5 4 3 2

Contents

v

Preface

Brickwork for Apprentices has been the basic reference book on brickwork for generations and a source of information for students/trainees starting out in his/her chosen career or a trained bricklayer seeking guidance in a particular area of study.

It has been an honour and a privilege to update this book for a fifth edition.

As technology advances with ever-increasing speed it has been necessary to produce this fifth edition to keep pace with the ever-changing programme of study required by today's students/trainees.

Some topic areas have been updated, extended or replaced, and new topics have been added.

Many of the topics covered are required for students/trainees who are studying for NVQs or CAs in brickwork but is equally useful for those studying for National and Higher National Certificates and GCSEs in construction.

The fifth edition contains 17 chapters which offer a structured programme of training and information.

Malcolm Thorpe

1 Craft training

Throughout the 1970s and 1980s there was great pressure for change in the way a craft skill is learned. Brickwork along with other construction industry trades has had its centuries-old tradition of Apprenticeship thoroughly examined.

There are three factors which have caused this re-examination of the Apprenticeship system, with a capital 'A':

  (i) A desire for retraining people who may wish to leave one industry and enter another

 (ii) A growing shortage of school leavers available and wanting to join construction trades throughout the 1980s, due to a falling birth rate 16 or so years earlier

(iii) A general feeling that just because you 'missed the boat' for vocational training when leaving school, you should not be denied the chance of learning a vocational occupation at any time in later life.

All this is far removed from the traditional arrangement of a school leaver joining a building company for a straightforward period of three or four years' apprenticeship, with attendance at a local college of technology, as the only way into the construction industry.

Very many patterns of vocational training have been proposed and tried, and continue to be developed at the present time.

Current ET (Employment Training schemes), in operation at college and training centres for adult learners, are the Government's response to change-factors (i) and (iii). Craft skills such as bricklaying, carpentry and plastering, which involve the use of tools and materials and require judgement of hand and eye, should not be confused with assembly processes.

The knowledge and practice required to understand how to assemble metal partitions or false ceilings are far less demanding than the skills of setting, cutting, shaping and finishing the materials of the bricklayer, carpenter and plasterer.

Building and construction's 'lead-industry body', responsible for developing change in training methods, has since 1964 been the CITB. This training board appreciates the difference between learning a craft skill in the one case, and that of becoming proficient at an assembly process in the second, by having separate development committees for the latter 'specialist subcontractors'.

Many operatives in the construction industry enjoyed a traditional Apprenticeship with a caring employer, where a training officer monitored site experience and also progress with further education at a college.

Others not so fortunate rather 'endured' their Apprenticeship, which has given the expression 'time serving' a somewhat hollow ring.

The long standing two-part C&GLI (City and Guilds of London Institute) Examinations of Craft and Advanced Craft Certificates based upon syllabuses of practical work and related technology have been phased out.

Traditional barriers to achieving a qualification such as length and method of training, where and when skills are acquired and the age of the student/trainee, have now been removed.

Objective measurement of a student/trainee's ability in the basic practical competences, step by step assessment of work modules, throughout the duration of a course of training aims to give a clearer indication of progress to trainer and trainee alike.

The construction industry is gradually moving towards a system where everybody working in it will require proof of their competence.

Schemes have been developed and are constantly being upgraded to keep in line with current trends: objective assessment of basic skills leading to NCVQ-approved (National Council for Vocational Qualifications), competence-based qualifications, which comprise a prescribed number of units of credit towards the award of an NVQ (National Vocational Qualification).

These, in association with C&GLI and CITB as the awarding body, are intended to fit in with the concepts of the EC (European Community), with whose policies of training for industry the UK is pledged to integrate.

Currently many competencies have to be achieved in the workplace to obtain an NVQ at levels 1, 2 and 3. This requires the student/trainee to be in employment and attend college part time. When students/ trainees are in full time education they follow a similar programme but achieve competencies in the workshop under simulation, therefore obtaining CAs (Construction Awards at Foundation, Intermediate and Advanced levels). These can be upgraded to NVQs when the student/trainee has found employment in the construction industry and evidence can be achieved in the workplace.

Whatever form learning a craft skill takes, however, it remains an apprenticeship with a small 'a'. Learning a craft remains a developmental process and must still provide sufficient time for repetition in practising the necessary manipulative skills of hand and eye on and off site, together with a sound knowledge of the technology of the trade, and the ability to draw if site plans are to be interpreted.

The student/trainee must realise that formal achievement of basic competences once only, in training, does not indicate total understanding.

Care must be taken by course organisers to see that sufficient job knowledge technology is retained in units of study leading to NVQs. Student/trainees need not only demonstrate how to carry out a craft operation but understand why it is constructed that way, if they are to gain the in-depth knowledge and ability to satisfy the demands of modern construction processes.

Despite all the changes in the manner and processes of learning a skill, the current uncertainties associated with training and the integration of C&GLI courses within the emerging structure of NVQs, brickwork remains an interesting, satisfying and challenging subject for a career.

2 Materials

It is a well-known saying in the industry that the craftsperson needs to understand the materials which they will use and lack of knowledge could result in materials being spoilt and work having to be taken down, both causing extra costs on the job.

Materials and methods are constantly being introduced into the industry and it is important that the users of these materials keep up to date with this ever-changing industry.

Bricks and their manufacture

The study of bricks, from raw materials to delivery of finished products, is an extensive one.

Being able to recognise a brick when it appears on site, know of its properties such as shape, size, weight, strength, porosity, colour etc. – and therefore know how and where to use it correctly – is all important basic knowledge.

Brick making is a very skilful business, with many individual variations in methods of manufacture between companies and their factories.

British Standards specify a brick as a walling unit designed to be laid in mortar and not more than 337.5 mm long, 225.0 mm wide and 112.5 mm high, as distinct from a building block which is explained as a unit having one or more of these dimensions larger than those quoted for bricks.

Bricks, which are one of the most durable materials, can be described as building units which are easily handled with one hand.

There are numerous uses for bricks but the main ones are as units laid in mortar to form walls and piers and the increasing use for brick paving.

Bricks were first made many thousands of years ago in hot climates, where a clay mixture was moulded and dried in the sun.

It was found that if the clay mixture was heated to a high temperature, the bricks were much stronger. The basic method of making bricks has not fundamentally changed.

Materials used for making bricks

Clay is the naturally occurring raw material used for producing most bricks. It consists mainly of silica and alumina.

Most clays also contain smaller amounts of limestone or chalk, iron oxide and magnesia. As natural deposits of clay in various parts of the country vary in their composition, a large variety of clay bricks are produced.

Suitable clays for brick making are reds, marls, gaults, loams, knotts and plastics, clay shales, refractory clays and brick earth.

They are found in many parts of the country, as shown in Table 2.1.

Table 2.1

Types	Areas used for brick making
Reds	Lancashire/Yorkshire
Marls	Cambridgeshire/Lincolnshire/Suffolk
Gaults	South East/East Anglia
Loams	
Knotts	Peterborough
Plastics	Central/East Anglia
Clay shales	Durham/Lancashire/Yorkshire
Refractory clay	Coalmining areas
Shale/brick earth	Kent/Essex

Note: Sometimes two or more types of clay are mixed together to produce bricks of varying colours and textures. Clays generally burn red, white or buff, according to the amount of metallic oxides they contain. Many different colours and shades have been created by the blending of clays.

Bricks are made by pressing a prepared clay sample into a mould, extracting the formed unit immediately and then heating it in order to sinter (partially vitrify) the clay.

Many types of brick may be produced, depending on the type of clay used, the moulding process and the firing process.

The stages of manufacture

There are six stages in brick manufacture, though many of these stages are independent. Figure 2.1 indicates the six stages that will take place.

1. Material excavation
The clay is excavated by machine from quarries close to the brickworks or brought into the brickworks from other quarries.

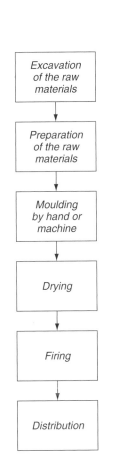

Figure 2.1 Stages in clay brick manufacture

Excavation of the raw materials

Preparation of the raw materials

Moulding by hand or machine

Drying

Firing

Distribution

2. Clay preparation

After collecting, the clay is prepared by crushing and/or grinding and mixing until it is of a uniform consistency. Some clays have to be weathered so that soluble salts are washed out of them. Water may be added to increase plasticity and in some cases chemicals may be added for specific purposes – for example, barium carbonate which reacts with soluble salts producing an insoluble product.

3. Moulding

The moulding technique is designed to suit the moisture content of the clay, the following methods being described in order of increasing moisture content.

(a) *Semi-dry process.* This process, which is used for the manufacture of fletton bricks (ex London bricks – now part of Hanson brick) utilises a moisture content in the region of 10%. The ground and screened material has a granular consistency which is still evident in fractured surfaces of the fired brick. The material is pressed into the mould in up to four stages. The faces of the brick may, after pressing, be textured or sandfaced.

(b) *Stiff-plastic process.* This utilises clays which are tempered to a moisture content of about 15%. A stiff plastic consistency is obtained, the clay being extruded and then compacted into a mould under high pressure. Many engineering bricks are made this way, the clay for these containing a relatively large quantity of iron oxide which helps promote fusion during firing.

(c) *The wirecut process.* The clay is tempered to about 20% moisture content and must be processed to form a homogeneous material. This is extruded to a size which allows for drying and firing shrinkage, and units are cut to the correct thickness by tensioned wires. Perforated bricks are made this way, the perforations being formed during the extrusion process. Wirecut bricks are easily recognised by the perforations or the 'drag marks' caused by the dragging of small clay particles under the wire.

4. Drying

The newly made bricks, known as 'green bricks', are dried slowly by standing them in the open or in drying chambers where waste heat is used. This helps to remove as much moisture as possible to prevent the formation of cracks, twists and warps, while the bricks are fired.

5. Methods of firing clay bricks

Apart from the pure white china-clay deposits, found only in Cornwall, most seams of clay and shale contain various impurities gathered over millions of years of geological time. Iron oxides are impurities which cause bricks to turn red or black during the firing process. Varying the kiln temperature, or the application of sand or other material to header and stretcher faces before firing, can produce single-colour brick types or result in multicolour facings.

Clamp burning. This is a centuries-old method of firing or burning stock bricks. There is no actual kiln structure. Coal dust fuel, approximately 5%, is mixed in with the clay during the tempering stage and before the bricks are moulded into shape.

The clamp is simply a solid stack of dried clay bricks built up upon a 400 mm thick layer of coke breeze fuel. There can be a million or more bricks in the clamp, which may have a simple corrugated iron roof structure over it to prevent rain from affecting the burning process.

The base layer of fuel is ignited via 'fire holes' at one end of the clamp, which slowly burns through over a period of weeks. The small percentage of coal dust within each brick contributes to the progression of the firing zone through the clamp.

Some manufacturers supplement the burning process with gas jets inserted along the clamp sides.

After burning through, the clamp of bricks is dismantled and the bricks sorted by eye into first and second quality, on the basis of hardness, colour and shape. Underfired bricks are used for covering and insulating the next clamp to be built.

Intermittent kilns. These are single chamber kiln structures of brick with walls approximately one metre thick, to reduce heat loss, loaded with dried green bricks; the entrance is temporarily bricked up with a wall built in lime mortar.

The temperature during the firing period of about 14 days is raised slowly at first to a maximum of approximately 1100°C and held for 7–10 hours. Gradual cooling over a period of days prevents cracks appearing as a result of rapid temperature change, before bricks are removed for inspection.

Hoffman kiln. This is really a terrace block of 20 or more intermittent kilns, all sharing one large chimney stack. Each separate chamber is loaded with about 20 000 green bricks and sealed, before hot air is admitted through flues in the party walls between chambers.

When a chamber has been pre-heated, the firing zone is concentrated there for 7–10 hours, with coal dust fuel added through small fire-holes in the flat roof. This type of kiln is used for fletton brick manufacture and permits a continuous cycle of operations for loading; pre-heating; firing; cooling; and unloading to take place around the total perimeter of the kiln.

Tunnel kilns. With all other types of kiln, the bricks remain in one place during the stages of drying, firing and cooling. With tunnel kilns, however, the bricks move through the stages of firing on trollies that run along steel rails laid through the length of the straight tunnel.

The maximum temperature firing zone, approximately half way along the tunnel, is fuelled by gas or oil fired jets operating from the flat roof of these kilns. Each 'kiln car' carries 2000 or more bricks and they are pushed slowly in a continuous train through the length of the tunnel. Kiln cars emerge after a total of 30 hours in the tunnel, to be steadily replaced by others loaded with more green bricks at the opposite end.

Tunnel kilns are installed in most new brick factories because of their accurate, computer controlled firing temperatures and general efficiency.

6. Delivery

Once the bricks have been checked they are usually banded and wrapped into packs of approximately 400 for delivery by either road or rail.

Hand-made bricks

These are made by 'throwing' clots of clay into a previously sanded mould. The excess clay is cut off with a wire and the 'green' brick turned out onto a pad ready for drying and firing. These bricks tend to vary in shape and size according to the amount of compaction of the thrown clay.

Classification of bricks

Bricks are classified according to their variety, quality and type. These classifications are independent of each other; examples of the same variety of brick may have different qualities.

BS 3921:1985 (Specification for Clay Bricks) provides manufacturers and architects with a more scientific classification of clay bricks and their properties than the descriptions given in Table 2.2.

Variety

Whether made from clay, sand and lime or concrete, bricks may be divided into four broad varieties: facings, commons, engineering bricks and refractories.

Facing bricks. These are made in a wide variety of colours and surface textures so as to be durable and attractive to look at.

Table 2.2 Some examples of traditional classification for clay bricks

Brick description	Classified by
Staffordshire blue engineering brick	Location Colour Use
Handmade red facing	Method of manufacture Colour Use
Solid wirecut common	Method of manufacture Use
Sandfaced fletton	Surface texture Method of manufacture
Perforated wirecut dragwire buff facing	Method of manufacture Surface texture Colour and use
London stock	Location Method of manufacture
Handmade sandfaced multi-colour Dorking stock	Method of manufacture Surface texture Colour Location

Commons. These are for general purpose walling which is most likely to be below ground level, externally rendered or internally plastered. They are not given particularly attractive surface features, but are hard and durable. Common bricks have been largely displaced by lightweight building blocks for internal partition walls and the inner leaves of cavity walling.

Engineering bricks. These are exceptionally hard, dense bricks which have a low porosity and therefore absorb very little water. Engineering bricks are intended for walls that are heavily loaded, or very exposed to risk of frost damage. They were originally developed by brick makers in Victorian times, in response to requests by civil engineers for a very strong brick for use in tunnels, bridges and viaducts.

Refractories. These bricks are made from specially selected clays which will withstand very high temperatures.

Quality
There are three qualities of brick:

Ordinary. Ordinary quality bricks are durable enough to be used in the external face of a building. They can resist frost attack and there is no limit on their soluble salt content.

Internal. Internal quality bricks need to be protected when used externally. There is no limit on their soluble salt content and they need not be frost resistant.

Special. Special quality bricks must be durable enough to withstand harsh weather conditions, where they will be constantly soaked with water and attacked by frost. They must have a limited soluble salt content.

Type
There are five main types of brick, these are shown in Fig. 2.2.

(a) Solid. The volume of pores in a solid brick must not be greater than 25% of the total volume of the brick.

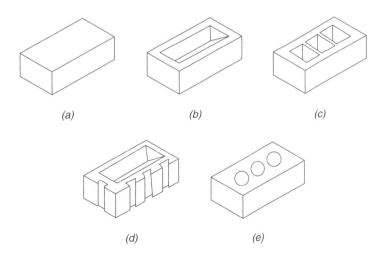

(a)   *(b)*   *(c)*

(d)   *(e)*

Figure 2.2 Types of bricks

(b) Frog. A frog is the depression formed in one or both bed faces of a brick. The volume of the frog should not be greater than 20% of the total brick volume.

(c) Hollow. The volume of the larger holes should be more than 25% of the total brick volume.

(d) Keyed. Dovetail grooves in one header and one stretcher face to provide a key for plaster or rendering.

(e) Perforated. The volume of the small holes will be greater than 25% of the total brick volume.

Indentations and perforations in bricks are shown in Fig. 2.3.

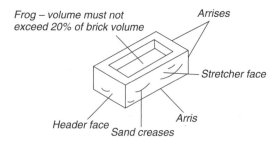

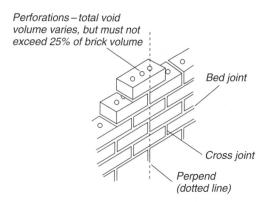

Figure 2.3 Definitions

Indentations and perforations may be provided for one or more of the following reasons:

1. They help in producing a strong bond between the bricks
2. They reduce the effective thickness of the brick and therefore reduce the firing time
3. They reduce the material cost of the brick.

Size of standard metric bricks

All clay brick manufacturers aim to produce their standard metric size bricks to these work size dimensions. Clay is a naturally occurring material, and variations in drying shrinkage mean that actual dimensions can vary from the intended work size; see Fig. 2.4.

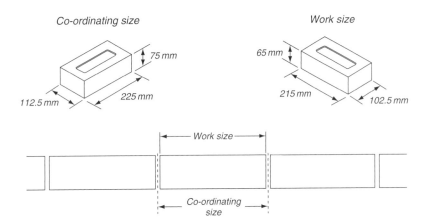

Co-ordinating size   Work size

75 mm   65 mm

225 mm   215 mm

112.5 mm   102.5 mm

Work size

Co-ordinating size

Figure 2.4 Sizes of standard metric bricks: co-ordinating (nominal) and work (actual)

A mortar joint all round each brick of approximately 10 mm allows brick-layers to stick rigidly to the co-ordinating size of 225 × 112.5 × 75 mm, by adjusting the mortar joint thickness as necessary.

Dimensional deviations
BS 3921 imposes limits on dimensional variations of clay bricks with a 24 brick test as illustrated in Fig. 2.5, using bricks taken at random from a delivery.

Compressive strength
Bricks may be specified by compressive strength depending upon where they are to be used in a building. Clay bricks can vary from 5 N/mm^2 (newtons per square millimetre) up to 100 N/mm^2 compressive strength.

Engineering bricks which have a compressive strength of not less than 50 N/mm^2 are called Class B, and may be either solid or perforated types. Those of more than 70 N/mm^2 are called Class A engineering bricks and are usually solid types. Average-strength facing bricks will have a compressive strength of about 20 N/mm^2. BS 3921 requires the crushing strengths of ten sample bricks to be averaged when testing for compressive strength.

Water absorption
Very dense engineering bricks may be used for a damp proof course (dpc), instead of flexible bitumen materials. Those bricks with only 4.5% by weight water absorption are suitable as a dpc in buildings. Those having not more than 7% absorption are adequate for a dpc in garden walls and other external works. BS 3921 requires the average of ten sample bricks to be tested when checking for water absorbency; see Table 2.3.

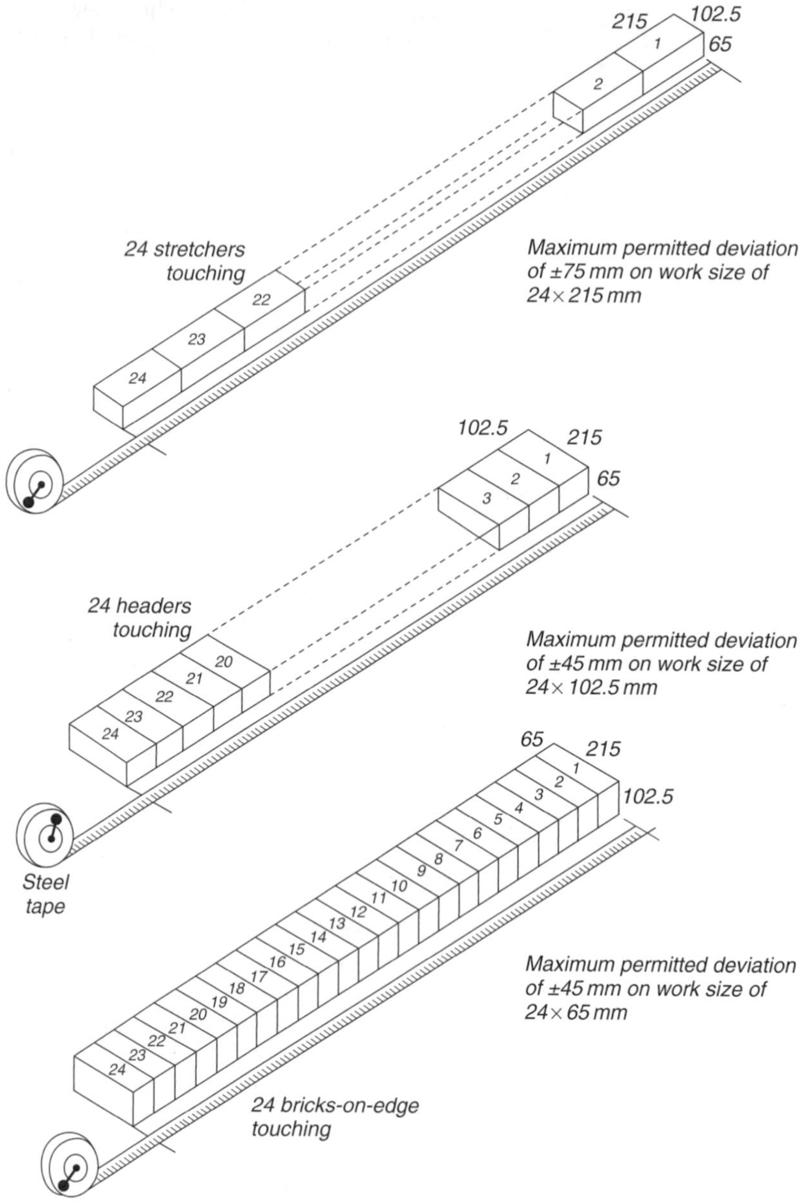

Figure 2.5 Carrying out a BS 3921 test for dimensional deviations

Frost resistance of bricks

Bricks are classified into degrees of resistance to frost, which is printed on the packaging of bricks as delivered to sites or otherwise indicated.

'F' Frost resistant designated bricks are durable in all building situations and can withstand freezing even when the wall is saturated.

'M' grade Moderately frost resistant bricks are durable except when in a saturated condition and exposed to freezing.

Table 2.3 BS 3921 classification of bricks

Class	Compressive strength in N/mm^2	Water absorption % by mass
Engineering A	Not less than 70	Not more than 4.5
Engineering B	Not less than 50	Not more than 7
DPC type 1	Not less than 5	Not more than 4.5
DPC type 2	Not less than 5	Not more than 7
All other bricks	Not less than 5	No limits are set

Bricks that do not meet with 'F' or 'M' grades are classed as '0'. These are not resistant to frost and need covering during construction and permanent protection from the weather, e.g. they are suitable for internal walls or behind cladding in a finished building. Manufacturers do not set out to make '0' grade bricks.

Soluble salt content in clay bricks
Other impurities in clay raw material from which bricks are made include various soluble salts. These salts are impossible to remove before bricks are made, and can cause problems of efflorescence and sulphate attack on cement mortar in walls which may be wet for long periods each year in completed buildings.

It is important that bricks with a low soluble salt content, 'L' grade, are specified for walls below ground level, retaining walls, parapets and chimney stacks.

Bricks given an 'N' grade rating, with normal soluble salt content, could be at risk only if used in walling exposed to continual dampness.

Durability of clay bricks
The two properties of frost resistance and soluble salt content are brought together in BS 3921, to give the architect/specifier an indication of the long-term durability of bricks, i.e. an 'FL' designation would indicate that the brick is resistant to frost and low in soluble salt content; see Table 2.4.

Table 2.4 Durability (BS 3921)

Brick designation	Frost resistance	Soluble salt content
FL	Frost resistant (F)	Low (L)
FN	Frost resistant (F)	Normal (N)
ML	Moderately frost resistant (M)	Low (L)
MN	Moderately frost resistant (M)	Normal (N)
OL	Not frost resistant (O)	Low (L)
ON	Not frost resistant (O)	Normal (N)

Other bricks

Calcium silicate bricks

More commonly called sand lime bricks on site, these are made from a carefully controlled mixture of 90% fine silica sand plus 10% lime. This mixture is pressed into steel moulds and then steam hardened.

Coarse flint sand is also used to make flint-lime bricks, which after steam curing have a higher compressive strength; see Table 2.5.

Table 2.5 Calcium silicate bricks – compressive strength classes

Class	Mean compressive strength not less than (N/mm²)	Colour marking of packs
7	48.5	Green
6	41.5	Blue
5	34.5	Yellow
4	27.5	Red
3	20.5	Black

Note: For comparisons, 1000 psi = 7 N/mm².

Loaded trolleys are pushed into steel steam chambers called autoclaves which hold a total of 22 000 bricks. Steam pressure is maintained for 7 hours to cause the sand and lime to fuse together. Upon cooling sample crushing tests allow the whole chamber to be given a compressive strength grading which may be between 21 N/mm² and 49 N/mm².

Colours
The basic colour of calcium silicate bricks, which are very smooth and of regular shape, is white. Different powder pigments added during the mixing stage of manufacture gives a range of facing brick colours. Multicolour and rustic surface texture calcium silicate bricks are also available.

Size
Calcium silicate bricks are manufactured to the same work size as clay bricks, namely $215 \times 102.5 \times 65$ mm, and may be solid or frogged. They contain no soluble salts and are frost resistant.

Compressive strength
Identical looking white calcium silicate bricks of different compressive strength batches are identified by paint colour marking of the packs as indicated in Table 2.5. The strength class of calcium silicate bricks, shown by number in Table 2.5, indicates pre-metric thousands of pounds per square inch (i.e. Class 4 indicated 4000 psi etc). (*Note*: 1000 psi is approximately equal to 7 N/mm².)

Concrete bricks

These are cast from a mixture of fine aggregate and Portland cement pressed into steel moulds. The natural setting and hardening process of cement determines the compressive strength of these bricks. They may be solid or frogged and are available as facings, commons and engineering quality.

Colour
The basic colour of concrete bricks is grey due to the Portland cement. Inorganic powder pigments are used to produce a range of plain and multi-colour bricks.

Surface texture
Concrete bricks are made smooth faced, exposed aggregate weathered, or split so as to resemble natural stone.

Size
The standard metric size is practically the same as for clay bricks, being $215 \times 103 \times 65$ mm. Metric modular bricks are 190 mm and 290 mm long to give courses 65 mm and 90 mm deep and are available to order.

Compressive strength
Facings and commons are made with a compressive strength of 21 N/mm^2: engineering quality at 40 N/mm^2.

Refractories

More commonly referred to as firebricks, these can be considered as two sorts:

(i) Very dense cream coloured solid bricks used to contain the fire in furnace linings, cement kilns, ships' boilers and for lining steel ladles transporting molten metal in a steel works
(ii) Very lightweight bricks used to 'back up' dense refractories, or to insulate chimney shafts in order to prevent heat escaping from flue gases. (This is the only brick that will float in water!)

Manufacture
Selected naturally occurring deposits of fireclay are moulded, dried and fired like other clay bricks to produce dense refractories able to withstand temperatures between 900 and 1350°C. Other fireclays containing a higher percentage of aluminium produce dense refractories able to withstand temperatures up to 2050°C.

The second sort of very lightweight refractories used for heat insulating purposes around the lower section of large chimneys are made from a special fireclay called diatomaceous earth.

Size
A standard size dense firebrick of $230 \times 114 \times 76$ mm is accompanied by a range of arch shapes, wedges and bullnose bricks made to suit circular section kilns and boilers. Cellular blocks of this same dense refractory material are used to protect the steel deck of the brick-transporter cars used in tunnel kilns.

Purpose-made bricks

Many buildings have decorative features which require shaped or moulded bricks. Manufacturers keep a standard stock of bricks for this purpose, some of which are illustrated in Fig. 2.6.

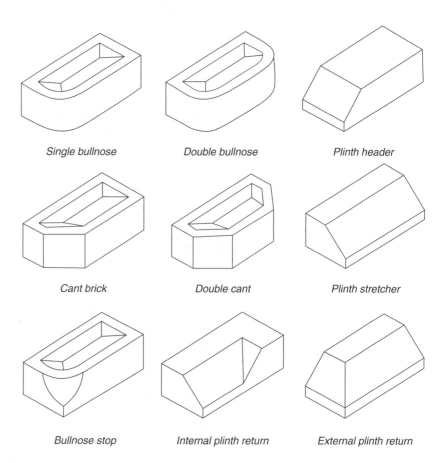

Single bullnose   *Double bullnose*   *Plinth header*

Cant brick   *Double cant*   *Plinth stretcher*

Bullnose stop   *Internal plinth return*   *External plinth return*

Figure 2.6 Types of purpose-made bricks

If an unusual shape is desired, the manufacturers will make these to the architect's specification. The advantages of using a purpose-made brick are that labour is saved in cutting, waste is avoided, and the natural surface is maintained.

Blocks

A block is described in BS 2028 as a walling unit exceeding the dimensions for bricks stated in BS 3921. Its height should not exceed either its length or six times its width.

Blocks are produced from clay and concrete.

Concrete blocks

Concrete blocks are produced in a range of shapes and sizes. The face side is usually 450×225 mm, the thickness varies from 37 mm up to 225 mm and the weight from 6.3 to 15 kg. They are produced in solid, hollow and multicut format. Multicuts enable a bolster cut to be made without wastage.

Special blocks

Special blocks such as the return block are usually designed to stiffen walls where bonding could cause weakness.

For closing cavities the reveal block could be used, another special is used to produce a splayed reveal.

Some manufacturers produce blocks with an insulant bonded to the outside face, others produce hollow blocks with an insulant inserted in the voids.

Walling built with pre-cast blocks may be divided into two main categories:

• Load bearing
• Non-load bearing.

Load bearing
These blocks are pre-cast in moulds and compacted with the aid of vibration, or moulding machines involving the use of compressed air, or a combination of both. These blocks are usually made of concrete comprised of Portland cement and a variety of aggregates, such as crushed stone, rock ballast, shingle etc.

Non-load bearing
These blocks can also be pre-cast in moulds but can also be produced in slab format and cut to size when set, and are usually made with cement and a variety of lightweight materials, such as fly ash, burnt coke etc.; see Table 2.7.

Foundation blocks

These are manufactured in widths from 250 to 335 mm, are used below ground level and are designed to support cavity walls. They may be dense or lightweight. The dense ones may require two persons to bed them.

The current Building Regulations state that hollow blocks must have an aggregate volume of not less than 50% of the total volume of the block calculated from its overall dimensions.

Hollow blocks must have a resistance to crushing of not less than 2.8 N/mm^2, if the blocks are to be used for the construction of a wall of a residential building having one or two storeys.

In all other circumstances blocks shall have a resistance to crushing of not less than 7 N/mm^2.

Block sizes

A block is described in BS 2028 as a walling unit exceeding the dimensions for bricks stated in BS 3921. Its height should not exceed either its length or six times its width.

Pre-cast concrete blocks are specified as Type A, Type B and Type C. Their sizes are listed in Table 2.6.

Table 2.6 Block sizes

Type of block	Length × height (mm)	Thickness (mm)
A	400 × 100	75, 90, 100, 140, 190
	400 × 200	140, 190
	450 × 225	75, 90, 100, 140, 190, 225
B	400 × 100 }	75, 90, 140, 190
	400 × 200	
	450 × 200	
	450 × 225	
	450 × 300 }	75, 90, 100, 140, 215
	600 × 200	
	600 × 225	
C	As above but intended for non-load bearing walls	As above but intended for non-load bearing walls

Clay blocks

The manufacture is similar to clay bricks using the extrusion/wirecut method. These blocks are made from finely washed clay with certain special properties, which is forced through an extruding machine, in the process of which the blocks are cut off to length as the continuous length of clay emerges from the machine.

The thickness of the walls of the blocks are about 12 mm and this allows the blocks to dry quickly and thoroughly.

The green clay blocks are then burnt at a high temperature.

The blocks are usually 300 mm long and 225 mm high and the thicknesses range from 37 to 100 mm for partition walls.

They are available with a smooth face and also dovetailed slots to provide a key for the plaster.

The clay blocks are not very easy to cut so the manufacturer produces special half units so that the bond can be formed without cutting any blocks.

Specials

Many block manufacturers provide specials cut to assist in the bonding on site and to prevent wastage from cutting; see Fig. 2.7.

REMEMBER – never use bricks and blocks in the same wall.

Table 2.7 Main types of blocks

Lightweight varieties	Internal non-load bearing walls and partitions generally	Breeze or clinker waste coke or ash and cement. Burnt clay	
Dense heavy	Internal load bearing walls and external walls	Usually concrete	
Hollow concrete	External walls, usually rendered	Concrete	
Cellular (lightweight)	Load bearing internal walls	Burnt clay	

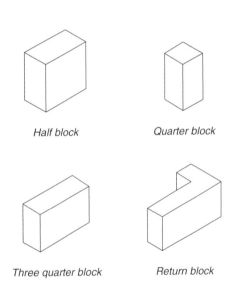

Half block   *Quarter block*

Three quarter block   *Return block*

Figure 2.7 Special blocks

Aggregates

Aggregates are divided into two main groups:

1. Fine aggregates – those aggregates which pass through a 5 mm sieve. These include the sands used for both concrete and mortar.
2. Coarse aggregates – those aggregates which are retained by a 5 mm sieve. These are used mainly in concrete production.

To produce good concrete, the aggregates must be sound, and of the type and quality specified.

Aggregates may be excavated from river beds or quarries or dredged from sand or shingle banks under the sea.

Aggregates form the bulk of mortar and concrete used throughout the construction industry.

They should be free from too much loam and clay and undergo certain tests before being accepted.

More detailed descriptions of the tests, and other tests for aggregates, are given in BS 812: methods for sampling and testing of mineral aggregates, sands and fillers, and it is recommended that a copy of this British Standard should always be available on site for reference.

Lime

Lime has been used very successfully for thousands of years as the only cementitious ingredient in mortars.

The lime used then was hydraulic; meaning that it was made from a naturally occurring chalk raw material which contains clay impurities. This clay content gave the lime a slow setting action not unlike the chemical hydration–setting–hardening process of ordinary Portland cement, but the lime mortar would take years to harden.

The raw materials used to produce building lime are either chalk or limestone.

When chalk or limestone is burnt at a very high temperature it turns into quicklime. This material cannot be used for building work in this state so it undergoes treatment by adding water to it – this process is known as slaking. The end product is hydrated lime.

Cement

If you read the writing on a cement bag you will see that its correct name is 'Ordinary Portland Cement'. This is the most commonly used cement on construction sites for making mortar and concrete. It is the all-important 'glue' in mortar which binds the grains of sand together when water is added and the setting and hardening process is completed. Used on its own, cement is too sticky, sets too hard and would develop severe shrinkage cracks. It is therefore always diluted with three, four or six equal volumes of sand.

The setting and early part of the hardening processes of Portland cement involve complicated chemical reactions between the mixing water and the cement powder in a batch of mortar or concrete. These reactions need to take place in damp conditions, called 'curing', if they are to be totally completed and give full hardened strength.

This early setting and hardening (see Fig. 2.8), taking place over hours and weeks respectively, must never be hurried by early 'drying out', as this will severely reduce the final strength of ordinary Portland cement mortar or concrete.

Figure 2.8 shows the stages in hydration of ordinary Portland cement as it sets and hardens, whether used in mortar or concrete.

Figure 2.8 Typical setting and hardening timescale for ordinary Portland cement

Mixing water added | Initial set not before 45 minutes | Final set not after 10 hours | Approx. ¾ of eventual strength is generated at 28 days | Hardening continues for many months

Plasticiser

Generally speaking bricklaying mortar made from cement and sand only is not sufficiently fatty or easily workable with the trowel. Such mortar is described as 'short' or 'harsh' and does not hold together when rolled on the spot board (see Fig. 8.1). Lime added to a mix improves the workability by temporarily retaining more mixing water, and results in a denser mortar. Lime is usually added in a volume equal to that of the cement in gauged mortar, i.e. 1:1:6. As an alternative, patent liquid plasticisers can be added, which generate millions of micro bubbles of air within the mortar as it turns in the mixer. Liquid plasticisers must be added by the mixer driver on site with great care – following the manufacturer's instructions, printed on the container. These are called air-entrained mortars.

Water

Mixing water triggers off a chemical reaction with cement, which causes the setting and hardening of mortar and concrete. Water should be clean enough to drink, as impurities can seriously delay or prevent this setting action.

Mortars

It is not usually difficult to cause an argument between bricklayers; ask them the following question. Does mortar stick bricks together or keep them apart? The answer is of course that it does both things, but a lot more besides. Mortar must stick firmly to bricks and blocks in external walling so as to keep the rain out. Mortar bed joints also hold bricks apart, so that the courses can be kept level and to an even vertical gauge of four courses to 300 mm, with standard metric bricks.

When walls were much thicker and Portland cement had not been invented, a mortar mix of lime and sand was used very successfully for thousands of years. The lime used then was hydraulic; meaning that it was made from a naturally occurring chalk raw material which contains clay impurities. This clay content gave the lime a slow setting action not unlike the chemical hydration–setting–hardening process of ordinary Portland cement, but the lime mortar would take years to harden.

Purpose

Bricklaying mortar is the ideal material for getting bricks to rest firmly upon each other, whether these are accurately shaped Class A clay engineering and calcium silicate bricks, or more irregularly shaped hand-made bricks.

The mortar must remain soft enough for each brick to be pressed down to the line, before suction causes the bed to stiffen up. Not only does mortar accommodate irregularities, but it must stick firmly to each brick so as to stop rain from penetrating exposed joints.

Types of mortar

The basic raw materials for bricklaying mortar can be prepared in a number of different ways, depending upon specification and site requirements.

The apprentice/trainee should understand the definitions given in Table 2.8.

Mortar designation groups
BS 5628:1985 divides bricklaying mortars into five groups for reference purposes, using Roman numbers (i) to (v) (see Table 2.9 and Fig. 2.9). The bricklayer section of a job specification or bill of quantities is just as likely to specify mortar for brick- or blocklaying by such a group designation number as it is to state mix proportions such as 1:1:6 or 1:1:5. Figure 2.10 shows that designation group (iii) mortar can be produced with cement/lime/sand, or masonry cement/sand or as a plasticised mortar, each having approximately similar compressive strength when hardened. The stronger mortar, designation group (ii), can similarly be produced in any one of these ways, but using less sand in each batch as indicated by the volume proportions.

Selection of mortar

As a general guide, the hardness or eventual compressive strength of mortar should be related to the hardness of, or preferably slightly weaker than, the bricks or blocks to be laid. So that if, as a result of slight foundation settlement, cracks develop, these will follow the joint lines, which can easily be cut out and re-pointed. Excessively hard mortar can result in the bricks becoming fractured at settlement cracks, thereby leading to a more extensive repair operation. Another reason for relating strength of mortar to the compressive strength of the bricks is to ensure that external walls weather evenly during the lifetime of the building, with any absorbed water evaporating at a similar rate from the surfaces of bricks and joints alike. In addition to being hard enough to transfer loads evenly between irregular surfaces of bricks, the choice of mortar must resist the effects of rain and frost in the long term; see Table 2.10.

Mixing mortar

Machine mixing is the most effective way of turning cement/lime and sand dry materials into mortar on site. A typical small tilling drum mixer is shown in Fig. 2.10. The mixer driver must be instructed in the importance

Table 2.8 Types of bricklaying mortar

Mortar type	Definition	Advantages	Disadvantages	Remarks
Lime mortar	Mortar made from 1 volume of lime to 3 equal volumes of sand	Smooth and workable. Ideal for training purposes where materials must be recycled	Unsuited to modern construction particularly cavity work; due to slow rate of hardening and final strength	1:3 mix using hydraulic lime, recommended by conservationists for repairs and repointing centuries-old brickwork
Cement mortar	Mortar made from 1 volume of OPC to 3 equal volumes of sand	Recommended only for use with Class A engineering bricks	Stiffens and sets too rapidly. Will cause shrinkage crack damage to other bricks and blocks if used by mistake	Usually has $\frac{1}{4}$ volume of lime added to improve workability
Compo	Composition mortar using separate site deliveries of dry OPC, lime powder in bags, and sand in bulk	Combines quicker strength gain of OPC with good workability of lime	Messy splitting of bags of cement and lime powder on site, giving health and safety risks	Obsolete for site use due to disadvantages given
Gauged mortar	Mortar prepared on site from bulk deliveries of wet-mixed LSM, to which 1 equal volume of OPC is added or 'gauged' each time a batch of mortar is required	Added powder pigments can be carefully measured and blended at works, before delivery to site, to ensure consistent colour of hardened mortar	Bulk delivery or stock-pile of LSM, to last a week or so, must be kept covered to prevent surface drying in summer and pigment blowing away. This, and rain washing, can weaken the colour strength of mortar batches	A reliable way of producing tinted bricklaying mortar
Masonry cement mortar	A purpose made cement for use only in bricklaying mortar. Mixed on site with 5 equal volumes of sand for general brickwork above dpc level	50 kg bags contain powder plasticiser ready mixed in with OPC powder. Does *not* require addition of lime or liquid plasticiser to batches of mortar	Masonry cement must *not* be used to make concrete	Unsuitable for concrete because it has been diluted with powder filler
Plasticised mortar	A cement/sand mortar to which a purpose-made liquid plasticiser is added at the site mixer to improve workability	A more convenient way to improve workability; easier to store and use than bagged lime	Excess mixing time will over-generate micro bubbles and result in a sloppy batch of mortar	Micro bubbles lubricate grains of cement and sand. Washing-up liquids must *not* be used instead of purpose-made plasticiser, as these can affect durability of mortar
Ready to use mortar	A complete RTU mortar delivered to site pre-mixed in plastic tubs. This CLM contains a chemical retarder which delays the setting action for 36 hours whilst in the tubs	No site mixer required. Weigh batching at supplier's depot removes problem of colour and strength variations between batches, which can arise with site mixed mortar	Difficulty of predicting exact daily requirements of RTU mortar, so as to avoid waste or bricklayer waiting time. Supply depots normally need a day's notice for deliveries	Full range of tinted RTU mortars available from suppliers in addition to natural colour

(*Abbreviations*: CLM – cement lime mortar, LSM – lime sand mortar, OPC – ordinary Portland cement, RTU – ready to use)

Table 2.9 Mortar designations with average compressive strengths

| BS 5628 Pt 3 Mortar designation number | Type of mortar | | | Average compressive strength in N/mm² at 28 days' site testing |
| | Air entrained mortars | | | |
	Cement/lime/sand composition mortar	Masonry cement/ sand mortar	Plasticised cement/sand mortar	
(i)	1:0 to ¼:3			11
(ii)	1:½:4 or 4½	1:2 or 3	1:3 or 4	4.5
(iii)	1:1:5 or 6	1:4 or 5	1:5 or 6	2.5
(iv)	1:2:8 or 9	1:5½ or 6½	1:7 or 8	1.0
(v)	1:3:10 or 12	1:6½ or 7	1:8	

Note: Numbers shown in the three middle columns indicate parts measured by volume the different mortar mixes. The symbol N/mm² indicates newtons per square millimetre stress.

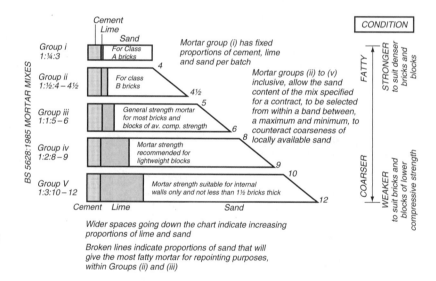

Figure 2.9 Cement/lime/sand mortar – another way of looking at mortar designations for brickwork and blockwork

of using gauge boxes of the correct size for every batch of mortar produced, and given the reasons why it is important to do so.

The description of this standard site mixer as a '5/3½' indicates the volume of dry materials put in and the volume of wet mortar discharged per batch (see Fig. 2.10).

Mixing water
The water added to each batch of mortar is not measured in the same way that an exact water/cement ratio controls what is added to a batch of concrete. A typical water/concrete ratio for concrete is 0.5. This means that

Gauged
mortar

Masonry cement
mortar

Plasticised
mortar

Size of gauge boxes shown in millimetres (internal dimensions), to give batches
of dry materials that will fill the mixer drum each time, allowing an extra 10%
of volume for BULKING of building sand and lime/sand pre-mix

Figure 2.10 Three ways of
producing a designation group
(iii) mortar

Total volume of
dry materials
put IN $0.14\,m^3$

Volume of wet-mixed
mortar OUT $0.1\,m^3$

5/3½
standard size tilting-drum
site mixer

Table 2.10 Mortar requirements

Good workability	Smooth and easy to handle with the trowel when transferring from spot board to wall, and applying cross-joints
Water retention	Mortar should contain sufficient fine particles of sand, together with cement and lime to prevent mixing water 'bleeding out' on the spot board
Adhesion	Mortar must stick to bricks and blocks to prevent rain penetration of joints
Durability	Mortar for externally exposed brickwork must resist the combined effects of rain, frost and any soluble sulphate salts in bricks of fired clay; see Table 2.4.

the amount of mixing water allowed is 50% of the weight of cement powder per batch. For example, with 50 kg cement, only 25 kg of water is
permitted (25 litres). Too much water weakens concrete.

Mixing mortar is a matter of what feels right: it must be workable with
the trowel: too dry and it will not spread into bed joints properly, too wet
and it smears the face of the bricks. The experienced mixer driver knows
the workability or consistency that bricklayers require, and he will soon
be told if it is coming out wrong!

Batching

The proportions of dry materials for each mix of mortar must be carefully and regularly measured/batched if the strength and colour of the hardened mortar is to be consistent. Variations can seriously affect durability of brickwork and result in conspicuous patchiness on completed facework elevations.

Bricks and mortar walls near the coast, for example, will need to be more resistant to the eroding effects of the weather than those sheltered by other buildings in towns and cities. Colour-matching or contrasting joints with the facing bricks, from the range of tinted mortar colours available, is another consideration.

Bulk deliveries of tinted lime/sand mortar, to which cement is added in carefully regulated proportions using a gauge box to produce 'gauged mortar', or daily deliveries of tinted 'ready-to-use' mortar, are both recommended methods.

Large-scale use of powder pigments added to the site mixer is not a reliable method, due to the difficulty of ensuring colour consistency of successive batches.

Measuring dry materials by the shovelful is totally unsatisfactory, as each will hold a different volume of cement powder or sand. Figure 2.11 shows

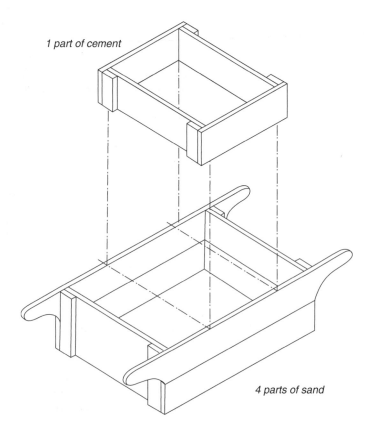

1 part of cement

4 parts of sand

Figure 2.11 Gauge boxes

typical bottomless gauge boxes, the regular use of which will ensure that measurement or batching remains consistent where mortar is mixed on site.

Placed on a flat surface behind the mixer, the larger is filled with sand and struck off level. The smaller, placed on the flat sand surface, is filled level with cement powder.

Both gauge boxes are lifted and the total contents transferred to the mixer by shovel. Figure 2.10 indicates the size of gauge boxes that will produce a designation group (iii) mortar which will fill the standard size mixer used on sites.

Retarded ready-to-use mortar

Figure 2.12 shows a typical delivery of retarded ready-to-use bricklaying/blocklaying mortar, direct to sites.

Figure 2.12 Delivery of retarded cement/lime/sand pre-mix mortar. Note the delivery day marked on each polythene cover to ensure use in rotation

The cement, lime and sand dry materials are mixed at a factory depot, complete with the required amount of water for optimum workability, plus a chemical retarder. This delays the onset of the initial setting action for 36 hours, while the mortar is in the delivery tubs. Upon delivery, these $0.3 \, m^3$ (600 kg) plastic tubs of mortar can be transported by fork-lift truck or tower crane directly to the point of use. Ready-to-use mortar avoids the risk of possible carelessness with site batching that can be the cause of colour and strength variations in finished brickwork. Bulk deliveries of lime/sand mortar plus bagged cement and a mechanical site mixer are all unnecessary if ready-to-use mortar is specified, a useful consideration on congested city centre sites.

Dry silo mortars

The modern approach on large mortar-using sites is to have a silo installed which is loaded with dry blended mortars that provide instant availability and consistent quality. This system requires space on the site to set up but reduces waste mortar as small quantities can be extracted. The silo has to be connected to water and power supplies.

The company providing the silo provides a top-up service. This system allows for increased consistency of the product whatever the environmental conditions, minimal wastage and there is no risk of contamination of the mortar. Any amount of mortar can be produced as required, uninterrupted by delays of any kind. The water content can also be adjusted to allow for the absorption of the various bricks and blocks.

Site requirements

A concrete pad is required at least 3 m × 3 m, constructed in an accessible position. The height of the silo is approximately 7.2 m and could weigh between 33 and 35 tonnes when full. The silo will be delivered and erected in position. It could be pre-loaded with up to 14 tonnes of dry mortar inside. A sensor is attached to the silo which tells the operator when there is only 10 tonnes of dry mortar remaining. Electrical and water supplies should also be available; see Fig. 2.13.

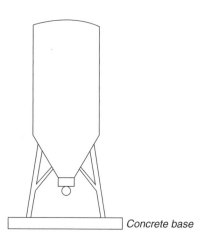

Concrete base

Figure 2.13 Dry mortar silo

Mortar testing

Increasingly, quality control procedures call for compressive strength tests to be carried out on a regular basis, throughout the bricklaying operations on site. The same steel moulds familiar for testing concrete are used to make 100 × 100 × 100 mm mortar cubes for laboratory testing at 7 day and 28 day intervals. The bricklayer section of a job specification will describe these mortar test requirements, and whether a copy of the mortar supplier's own test results will be acceptable, or if an independent laboratory is to be appointed.

Typical compressive strengths of different mortar mixes are given in Table 2.9. These are results based on site tests, and will show a continuing slight increase in the weeks following a 28 day test.

It is worth noting that the typical compressive strength of ready-to-use mortar is ultimately 25–30% higher, due to better batching control by weight rather than volume, and also because the slowed rate of initial setting and hardening improves hydration and eventual strength gain of the cement in the mortar.

Sulphate attack on Portland cement mortars

The use of sulphate resisting cement prevents sulphate attack from developing in any brickwork that may be saturated for long periods of time (see Table 2.11). Sulphate resisting cement will therefore be required for brickwork below dpc level in those subsoils which contain water soluble sulphates.

Table 2.11 General recommendations for mortar mixes

Use	BS 5628 Mortar designation number	Types of mortar		
		Cement/lime/sand mortar	Masonry cement/sand mortar	Plasticised cement/sand mortar
		(parts by volume)		
Class A and B engineering quality brickwork	(i)	1:0 to ¼:3	–	–
Work up to dpc level, including extremely exposed brickwork, e.g. chimney stacks, parapets, free-standing boundary walls, BOE sills and copings	(ii)	1:½:4	1:3	1:4
External facing and common brickwork above dpc level and internal walls of medium density blockwork	(iii)	1:1:6	1:4	1:5
Internal walls of lightweight blockwork	(iv)	1:2:8	1:6	1:7

Chimney stacks, parapets and free-standing boundary walls constructed of fired clay bricks, manufactured from a brick-earth that contains soluble sulphates, will also be at risk unless sulphate resisting cement is specified.

Reinforced mortars

This is a term used to describe those admixtures such as sytrene butadiene, added to cement/lime mortars to improve adhesion and water resisting properties of brick-on-edge sills and copings. Great care must be exercised to follow the manufacturer's instructions carefully if these admixtures are to work effectively and be used safely.

Concrete

Concrete is an artificial rock made from a mixture of coarse aggregates, sand, a cement binder and water. It is one of the few building materials that can be produced on the building site. Other materials added at the mixer are referred to as 'admixtures'.

The appearance and properties of concrete are similar to those of limestone rock. The main advantage of using concrete is its versatility. It can be moulded to any required shape and its load bearing capabilities can be increased by casting in steel reinforcing bars.

There are several types of concrete.

- Dense concrete
- Lightweight concrete
- Air entrained concrete.

Constituents

The three main constituents used to manufacture concrete are:

- Cement
- Aggregates
- Water.

Manufacture of concrete

The manufacture of hardened concrete involves two stages. These are the plastic (setting) stage and the rigid (hardening) stage.

During both of these stages the chemical process of hydration occurs, where the cement reacts with the water.

The aggregate, although present, does not take part in the chemical reaction.

$$\text{cement} + \text{aggregate} + \text{water} > \text{concrete} + \text{heat}$$

Plastic stage

1. *Mixing.* The constituents are mechanically mixed together in the correct proportions to give a homogeneous (same consistency throughout) concrete mixture. During mixing, the cement and water produce a paste, and a film is formed around each aggregate particle. The finer aggregate particles fill the voids between the coarse aggregate.
2. *Placing.* The concrete mixture is placed into a mould to obtain the required shape.
3. *Compaction.* The concrete mixture may need to be vibrated to remove any air voids formed during placing. This is known as compaction of the concrete mixture.

Rigid stage

When the concrete mixture has set in the mould, the hardening process starts.

Curing. This is the process of retaining water in the concrete mix and maintaining the temperature of the concrete at about 20°C, which ensures that the cement binds the aggregate particles together and that the concrete hardens at a favourable rate. Curing is carried out by protecting the concrete from the weather. The exposed concrete surface is covered with a water resistant material such as plastic or with damp canvas or hessian. This stops evaporation.

If the temperature drops below 5°C, the hardening process almost stops, and if the temperature is too high, the temperature difference between the concrete and the surroundings can cause cracking.

The concrete mix. Two essential properties of hardened concrete are durability and strength. Both properties are affected by the voids or capillaries in the concrete which are caused by incomplete compaction or by excessive water in the mix.

It is important in the manufacture of concrete to mix the aggregates, cement and water in the correct proportions, in order to obtain the correct workability and strength required for the job.

A typical mix contains the following proportions of ingredients:

1 part cement : 2 parts of fine aggregate : 4 parts of coarse aggregate : 0.5 parts of water

Type of mix. The proportions of the concrete may be mixed by one of two methods:

• Mixing by volume
• Mixing by mass.

Mixing by volume If the proportions of the ingredients are measured by volume, the mix obtained is known as a 'nominal mix'. Specified mixes are based on the aggregate and the cement being dry when measured.

Specified examples of nominal mixes:

• Mass concrete 1:3:6
• Reinforced concrete 1:2:4.

A minimum of cement is used in mass concrete to reduce cost. More cement is used in reinforced concrete to increase workability. In nominal mixes there is always a 1:2 ratio of fine aggregate to coarse aggregate. This is to ensure there is enough fine aggregate to fill voids between coarse aggregate particles. The amount of water has not been specified but will be found from tests carried out on site as described. Mixing by volume has poor control. This means that there is poor control over the proportions of ingredients because mixing by volume does not take into account the water content of the aggregate. Mixing by volume is not very often used these days.

Mixing by mass For any concrete mix, the mass of each of the ingredients needs to be known accurately. These are called the batch quantities. An accurate knowledge of the batch quantities enables the properties of the concrete to be forecast to a good precision.

The properties of concrete are today much better understood.

There are many varieties of concrete. This is because:

1. the ingredients can be varied;
2. the relative proportions of the ingredients can also be varied.

The properties of the concrete can be greatly modified by these two methods. BS 5328 shows several grades of concrete, their compressive strengths at 28 days and the permitted applications at these strengths; see Table 2.12.

Table 2.12

Grade	Characteristic compressive strength at 28 days (N/mm²)	Lowest grade use	Minimum cement content (kg/m³)
C7.5	7.5	Plain concrete	120
C15	15.0	Reinforced concrete with lightweight concrete	180, 240*
C20	20.0	Reinforced concrete with dense concrete	220, 240*

*When used for reinforced concrete.

Workability. The measurement of workability of fresh concrete is of importance in assessing the practicability of placing and compacting the mix and also in maintaining consistency throughout the job.

In addition workability tests can be used as an indirect check on the water content and therefore on the water/cement ratio of the concrete.

3 Tools

A craftsperson is judged by his/her tools. The saying 'Clean tools, clean job' applies to all crafts, and bricklaying is no exception. The apprentice will never regret the small amount of time and trouble spent in cleaning tools at the end of a day's work. All that he/she requires for this are an old rag and a piece of brick for scouring.

Tool kit

Bricklayers can carry their complete tool kit in a medium size canvas shoulder bag or heavy duty holdall in one hand, and move about site with a spirit level in the other. A kit of tools is very personal property and the apprentice will become familiar with all the items, particularly the brick-laying trowel, and will not want other bricklayers to use them.

A tool kit must be maintained in good order so as to ensure efficient and safe working, avoiding accident to self and others. On construction sites safety helmets must be worn at all times for brain protection – we only have one! Eye protection must be worn when cutting brickwork, blocks or concrete to protect irreplaceable eyes – we only have two!! Feet should be protected with stout safety shoes or boots. Soft top track shoes give no protection against a hammer accidentally dropped from hand height, never mind anything worse. Do not throw your tool bag down, as this will damage and loosen handles.

Brick trowel

This is the most heavily worked item in a bricklayer's tool kit, used for gathering and spreading mortar, and for rough cutting some kinds of brick. Available in a range of shapes, sizes and thickness of steel, with length of blade from 230 mm to 330 mm. Choice of trowel has a lot to do with personal preference and what feels most comfortable for the individual.

The apprentice should avoid the temptation that biggest is best.

Narrow blade 'London pattern' trowels are suited to cavity work, broad heel trowels lend themselves to solid walling. The best trowels are solid forged from a single piece of steel from tip to tang.

The majority have one side of the blade of thicker steel to withstand wear and tear of tapping bricks and rough cutting. For this reason, left and right handed versions are produced. Philadelphia pattern brick trowels are not intended to be used for rough cutting, and so are not 'handed'. Use of the trowel handle to tap bricks down to the line should be avoided as this tends to loosen the handle and 'burr' the end, making it uncomfortable to use.

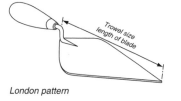

London pattern

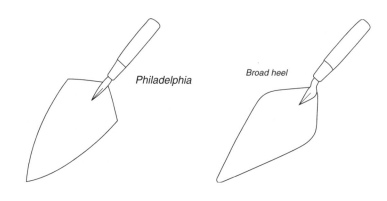

Philadelphia   *Broad heel*

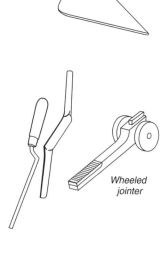

Pointing trowels

These are made with blade lengths from 75 mm up to 175 mm. That with the shortest blade, sometimes referred to as a 'dotter', is used for filling and striking cross joints.

The longer blade pointing trowels are used with a hand hawk for filling and striking bed joints.

Jointing tools

A variety of tools are shown for applying a permanent finish to the exposed surface of mortar joints, some purpose made, some produced by the bricklayer.

Patent wheeled jointers are ideal for raking out mortar to a constant depth in preparation for square recessed joint finish.

Wheeled jointer

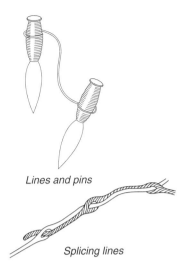

Lines and pins

Splicing lines

Line and pins

No tool kit is complete without a 'set of lines' for controlling level, line, plumb and gauge of any walls over 1.200 m long. Always buy the best quality hardened steel pins you can afford so that they last.

Bricklayer's line made from traditional hemp can be spliced if accidentally broken so as to avoid knots. Cotton, polyester and various types of nylon are also sold, but it is impossible to join any break in these materials by splicing, as the strands do not separate neatly. The finer the line the better for accurate work. When renewing line, wrap insulating tape around the pins first to prevent rust staining. Always tie the line-end on to pins before winding on, in case you drop the pin when working on a high scaffold.

Club or lump hammer

One kilogram size is ideal for use with a bolster for fair cutting and also for cutting away existing brickwork with a cold chisel.

These have ash- or hickory-wood handles or steel with a rubber sleeve.

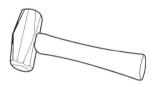

Brick hammer

Used for rough cutting very hard bricks which would damage a trowel.

They have one hammer head and one forged chisel end which can be reground when worn.

Bolster chisels

A 100 mm width blade is most useful for fair cutting bricks, and bolsters can be supplied with or without rubber or plastic collars for hand protection.

Bolsters must be kept sharp for efficient operation.

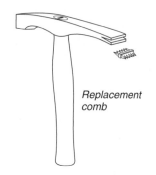

Replacement comb

Comb hammer

This has a hammer head one side and is slotted to take hardened steel replacement combs at the other side.

Scutch

This is slotted both sides to receive renewable hardened steel combs or blades.

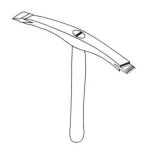

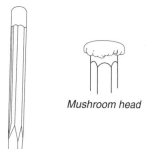

Mushroom head

Cold chisels

These must be kept sharp if cutting away brickwork is to be as painless as possible. Grind off the first beginnings of burring over, don't wait until a full grown and dangerous 'mushroom' has developed. Always use a steel point and eye protection if you are unfortunate enough to be cutting away concrete, don't ruin a cold chisel.

Comb chisel

A very useful substitute for a cold chisel when cutting away brickwork.

Plugging chisel

A purpose-made tool for carefully cutting and toothing out joints in existing brickwork.

Measuring tapes and rules

Folding boxwood or plastic rules tend to get broken easily and have been largely replaced by steel tapes.

These also have a limited life as they tend to get full of grit and do not fully retract.

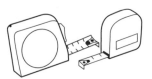

Hand hawk

Allows mortar to be picked up conveniently with a pointing trowel.
   The removable handle makes it easier to fit into a tool bag.

Hand brush

Medium to fine bristle is used for lightly brushing face brickwork at the end of a day's work. Great care must be taken not to leave bristle marks in any mortar still soft.

Small square

For pencil marking bricks accurately before cutting.

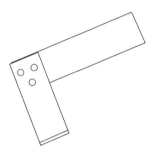

Dividers

For spacing out voussoir positions on an arch support centre (see Chapter 9) either side of the key brick location before starting work.

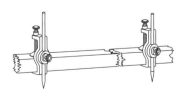

Bricklayer's sliding bevel

Used for transferring and marking the same angle of cut for all the bricks when raking cutting to a gable-end wall or when carrying out tumbling-in.

Pair of trammel heads

One a steel compass point, the other for holding a pencil. For use with a timber lath as a beam compass, when making a template for constructing a curved wall, drawing arches full size or striking the shape of the brick core to a bullseye.

Pencil

For plumbing perpends on face brickwork, H or 2H grade pencils will last longer than HB grade.

Masonry hand saw

With tungsten carbide tipped teeth for toothing out existing brick- or blockwork where the mortar is not too hard, and also for neatly cutting through lightweight aerated concrete blocks.

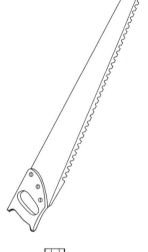

Retractable blade knife

Very useful for cutting dpc material and sharpening pencils.

Spirit levels

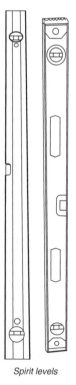

These have a dual purpose and are used for checking the horizontal and vertical accuracy of brickwork.

For general-purpose use, a bricklayer's spirit level should be not less than one metre in length, preferably 1200 mm long.

Available as girder section hardened aluminium with hand-hold slots, or as a hollow box section enamelled aluminium, or in hardwood.

A professional bricklayer never taps or knocks a spirit level when plumbing or levelling brickwork, 600 mm long spirit levels are made for use in restricted places, 200 mm long boat levels are made for plumbing soldier bricks and for decorative brickwork.

Checking the spirit level
It is important to remember that some spirit levels are adjustable but the cheaper ones are not.

Great care should be taken of the spirit level as it is expensive and if ill used can lead to inaccurate levelling and plumbing.

Checking for level
It is therefore important to check the spirit level occasionally to ensure its accuracy.

If a course of bricks, a sill or a lintel has been set perfectly horizontal, 'reversing the level' end-for-end should confirm this; see Fig. 3.1.

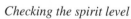
Spirit levels

Figure 3.1 Daily spirit-level check: horizontal level

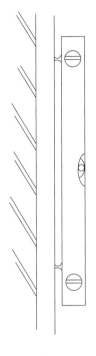

If the spirit bubble reads truly horizontal in Fig. 3.1, but is not horizontal when the level is reversed, then the spirit level is out and needs to be adjusted.

The clamping screws for the horizontal bubble tube must be slackened, and the necessary adjustment made.

Checking for plumb
Set two screws equal to the length of the level apart in a vertical position on a door frame or jamb.

Check that they are plumb with a plumb bob or with a spirit level that is known to be accurate.

Position a faulty level onto the screws and adjust until the spirit level reads plumb.

Reverse the level and adjust if required.

This procedure needs to be repeated if the level has double bubbles.

Once these positions have been produced they could be used for all tradespersons requiring levels checking.

Items of equipment additional to the tool kit

These may be made up by the bricklayer.

Brick cutting gauge

For quick accurate marking of standard size cut bricks required for bonding purposes. This item should be made from oak or other hardwood in order to survive the rigours of life in a tool bag.

Corner blocks

A simple means of supporting bricklayer's line at quoins that do not leave pinholes behind. If made from hardwood they will last longer.

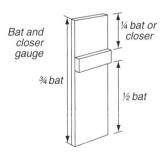

Bat and closer gauge

¼ bat or closer

¾ bat

½ bat

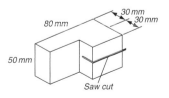

Hardwood corner block

Frenchman

A simple device made from an old table knife, heated, bent over and filed to shape for trimming mortar bed joints when carrying out weather-struck and cut pointing, or tuck pointing.

Storey rod

Indispensable for building and checking courses of brickwork and block-work, so as to ensure consistent vertical gauge above and below datum levels.

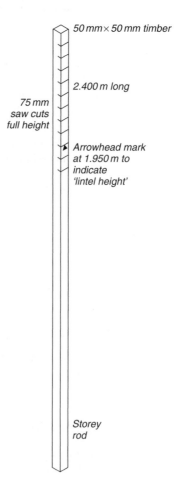

50 mm × 50 mm timber

2.400 m long

75 mm
saw cuts
full height

Arrowhead mark
at 1.950 m to
indicate
'lintel height'

Storey
rod

Feather edge pointing rule

Used in conjunction with a Frenchman for trimming bed joints after pointing. Cork pads allow trimmed mortar to fall away cleanly.

Maintenance of tools

It is important to take care of tools and equipment to ensure that they are clean and sharp when required.

Mortar should never be allowed to harden on the blade of trowels as this creates a rough surface and prevents free and easy movements when picking up and spreading mortar.

All cutting tools should be cleaned after use. Never knock bricks or blocks with the end of the handle as this will damage it. Have the chisel end sharpened and tempered regularly. Skutch hammers need cleaning after use and the combs replaced as necessary. Check all handles for cracks and splits and change where necessary.

All chisels need to be kept sharp and regular attention should be paid to the head to prevent mushrooming. Plastic mushroom sleeves are available for most chisels to reduce the risk of injury to the hand holding the chisel.

Levels as previously mentioned should be checked regularly for level and plumb. They should be kept clean, especially the glass/plastic covering the bubbles.

Lines need constant attention to prevent twisting, fraying and stretching. Avoid knocking line pins into hardened joints as this will bend the point of the line pins.

Safety

Eyes

Remember to have personal protective equipment ready when required. Always use goggles when cutting.

Heads

Since 1990 everyone on a construction site has to wear a helmet by law. These have to be renewed when out of date.

Feet

Wear safety boots to support ankles and protect the feet. If Wellingtons are required ensure these are also protected.

Hearing

Wear ear protectors when using items of plant and equipment which emit noise over the recommended level.

Lungs

Wear masks where dust is produced.

4 Bonding of brickwork

The brick is a relatively small unit and it can therefore be manipulated to fit most dimensions.

To achieve economy in both materials and labour the architect will give thought to the planning of these dimensions while the bricklayer will exercise skill in setting out. Whatever the system of measurement adopted bonding principles must still apply.

Brick proportions

The apprentice should note that in the following bonding examples the standard brick format is used, namely $225 \times 112.5 \times 75$ mm.

Using a 10 mm mortar joint the actual size of the standard brick becomes $215 \times 102.5 \times 65$ mm. Figure 4.1 indicates the basic relationship between header and stretcher faces with joint allowances for standard bricks. The cut pieces of whole bricks are used in bonding.

The majority of brickwork produced is in the form of cavity wall construction for the external walling of buildings (see Chapter 10). Stretcher bond makes the most economic use of expensive facing bricks in the half-brick thick (102.5 mm) outer leaf of this type of construction; see Fig. 4.2(a).

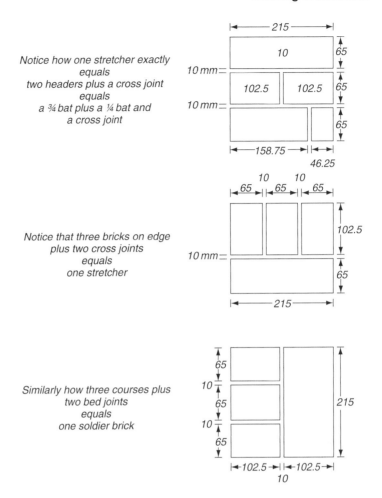

Notice how one stretcher exactly
equals
two headers plus a cross joint
equals
a ¾ bat plus a ¼ bat and
a cross joint

Notice that three bricks on edge
plus two cross joints
equals
one stretcher

Similarly how three courses plus
two bed joints
equals
one soldier brick

Figure 4.1 Brick proportions

It is, however, most important for the apprentice to have a wider understanding of the basic principles and rules of bonding which are best illustrated on solid walling greater than 102.5 mm thick. The apprentice will encounter non-cavity or solid brickwork in freestanding boundary walls, building refurbishment, earth retaining walls, chimney breasts, civil engineering brickwork, inspection chambers etc.; see Figs 4.2(b) and 4.2(c).

Purposes of bonding

The reasons for bonding brickwork are:

(i) to strengthen a wall;
(ii) to ensure that any loads are distributed; and
(iii) to make sure that it is able to resist sideways or lateral pressure; see Fig. 4.3.

The straight joints of an unbonded wall make it weak and liable to failure, as shown in Fig. 4.4.

(a)

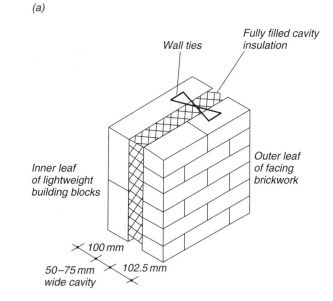

Wall ties

Fully filled cavity insulation

Inner leaf of lightweight building blocks

Outer leaf of facing brickwork

100 mm

50–75 mm wide cavity

102.5 mm

(b)   *(c)*

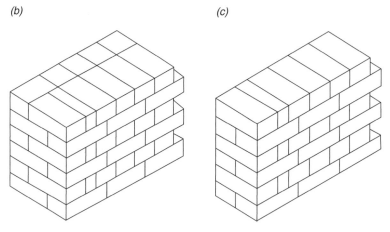

Figure 4.2 (a) A typical external cavity wall. (b) A typical one and a half brick wall. (c) A typical one brick wall

Load

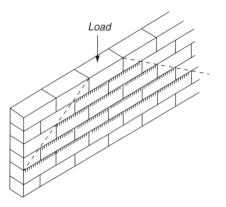

Figure 4.3 Bonding the bricks distributes any applied loads evenly

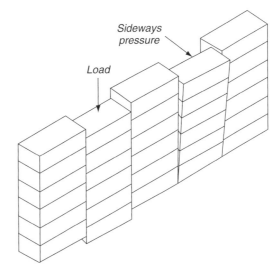

Figure 4.4 Tendency to failure of unbonded walls

Principles of bonding

To maintain strength, bricks must be lapped one over the other in successive courses along the wall and in its thickness.

There are two practical methods, using either a half-brick lap or a quarter-brick lap, called half-bond and quarter-bond (Fig. 4.5).

If the lap is greater or smaller than these, then both appearance and strength are affected. If bricks are so placed that no lap occurs, then the cross-joints or perpends are directly over each other (Fig. 4.6), and this is

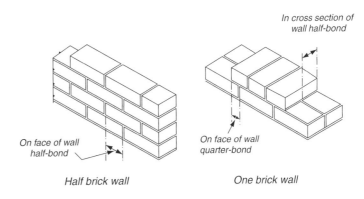

Figure 4.5

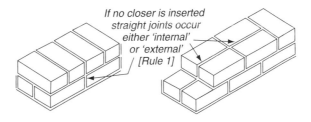

Figure 4.6

termed a 'straight joint', being either 'external' for those appearing on the face of the wall, or 'internal' for those occurring inside the wall, and they should be avoided whenever possible.

The apprentice should note that internal straight joints will occur in some bonding problems; excessive cutting would perhaps solve a particular problem, but this wastes labour and materials and tends to weaken the wall. On the other hand, by introducing one or two straight internal joints, whole bricks can be used. This is a case where practice and theory must compromise.

The pattern in a brick wall is purposely arranged, has its particular use, and is called a 'bond'.

To summarise, the two main principles of the bonding of brickwork are:

1. To maintain half- or quarter-bond, avoiding at all times external straight joints and internal straight joints wherever possible.
2. To show the maximum amount of specified face bond pattern as possible.

To assist in maintaining these principles, rules should be remembered and applied (see Fig. 4.10).

The apprentice should never try to remember all the problems shown as examples. Problems must be solved as they occur by the logical application of the Rules. Eventually, the bonding of brickwork becomes automatic to the bricklayer.

Several bonds are in general use, but for the purpose of beginning the apprentice's bonding education, 'stretcher', 'English', and 'Flemish' bonds will be explained. Problems in other bonds can be solved by the application of the same rules.

Note. Wall thicknesses are usually stated in brick sizes, e.g. the width of a brick is known as a half-brick wall; the length of a brick as a one-brick wall; the width, plus length of a brick, as a one-and-a-half brick wall, and so on.

Stretcher bond: used in the building of half-brick walls. The face bonding consists entirely of stretchers, except where return angles, stopped ends, and cross walls occur. To gain maximum strength, half-bond must be maintained at all times. At the junction between two walls, quarter-bond is introduced, but by the insertion of three-quarter bats the wall can be continued in half-bond (Fig. 4.7).

English bond: alternate courses of headers and stretchers. A very strong bond, with no straight joints occurring in any part of the wall. Being monotonous in appearance, it is used in walls where strength is preferable to appearance; see Fig. 4.8.

Flemish bond: alternate headers and stretchers in the same course. Used in brick walls of a decorative nature. Internal straight joints, quarter-brick in length, occur at 100 mm intervals along the middle of the wall. The header must be in the centre of stretchers in courses above and below (Fig. 4.9).

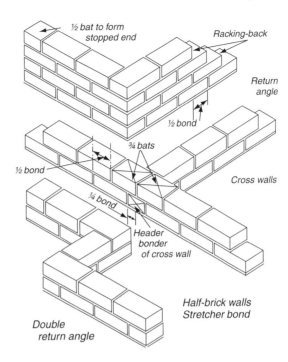

Figure 4.7 Junctions in half brick walls

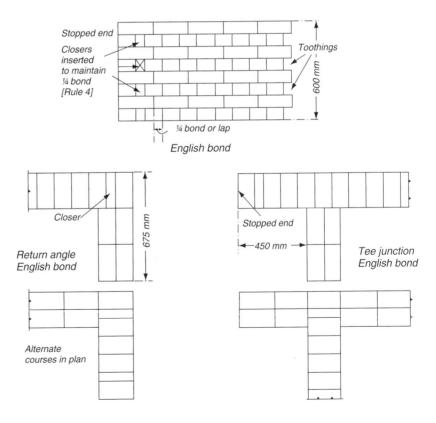

Figure 4.8 English bond

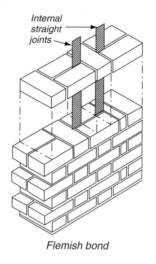

Flemish bond

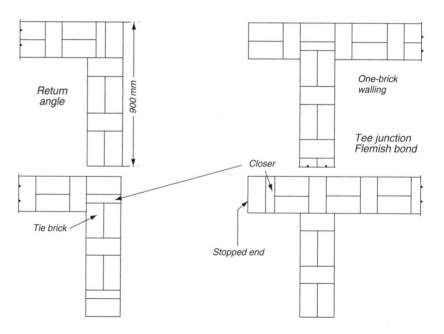

Figure 4.9 Flemish bond

Rules of bonding	
Rule 1 The LAP of brickwork along the face of a wall shall be ¼ brick 56.25 mm. Exception: Half lap of 112.5 mm in stretcher bond.	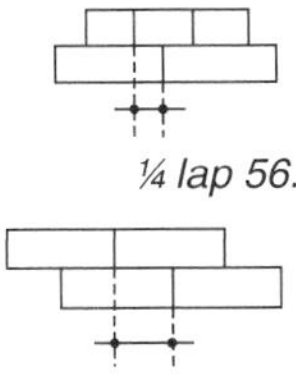 ¼ lap 56.25 mm ½ lap 112.5 mm
Rule 2 The LAP of brickwork across the width of a wall shall be 112.5 mm.	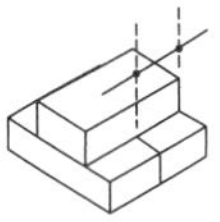
Rule 3 Brickwork should be set out on the face side and from each end of a wall or pier, so that any BROKEN BOND is centrally located. Exception: Reverse bond, where end bricks do not correspond.	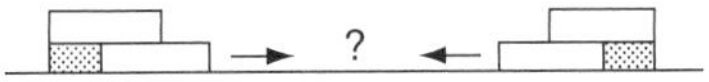 ?
Rule 4 In English and Flemish bonds a queen closer must be placed next to the QUOIN HEADER, in order to establish ¼ lap. Exception: It is permissible to use a ¼ brick instead in Flemish bond, if this will avoid broken bond at centre, or is otherwise preferred.	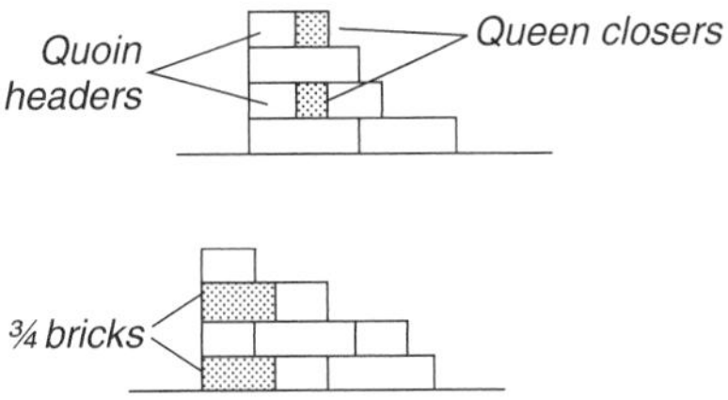 Quoin headers — Queen closers ¾ bricks
Rule 5 Cross joints in consecutive courses of bricks must not coincide one above the other to create STRAIGHT JOINTS.	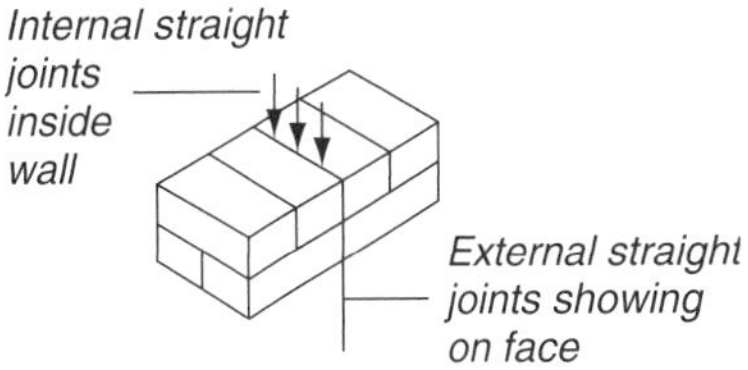 Internal straight joints inside wall External straight joints showing on face
Rule 6 Cross joints in alternate courses must coincide vertically one above the other on the face of the wall. PERPENDS, see Fig 4.17.	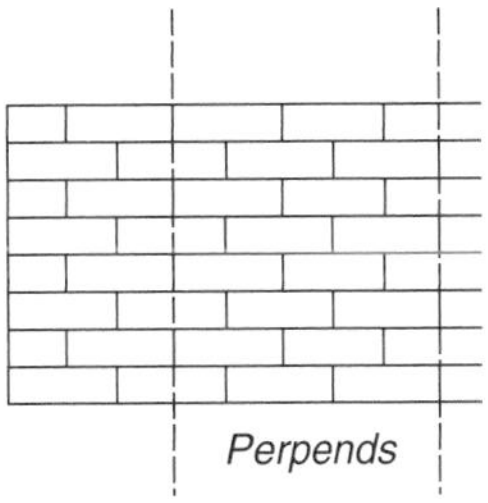 Perpends

Figure 4.10 Rules of bonding

Rule 7 *Cross joints shall pass through a wall unless stopped by the middle of stretcher, to produce SECTIONAL BOND. (See Figs 4.11, 4.12 and 4.13.)*	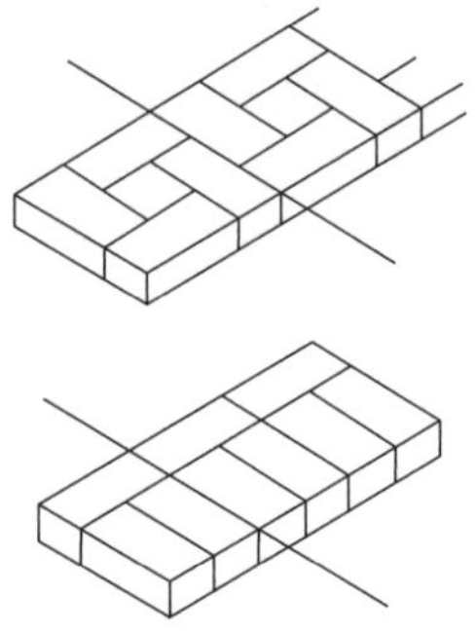
Rule 8 *The TIE-BRICK between one brick thick English bond walls is always a HEADER.*	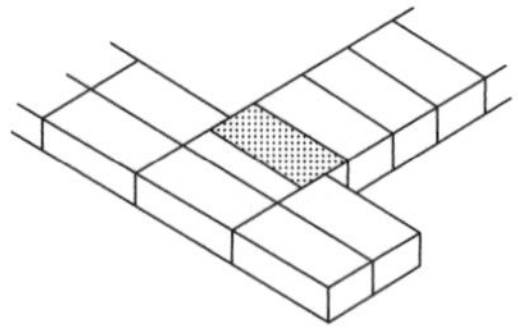
Rule 9 *The TIE-BRICK between one brick thick Flemish bond walls is always a STRETCHER.*	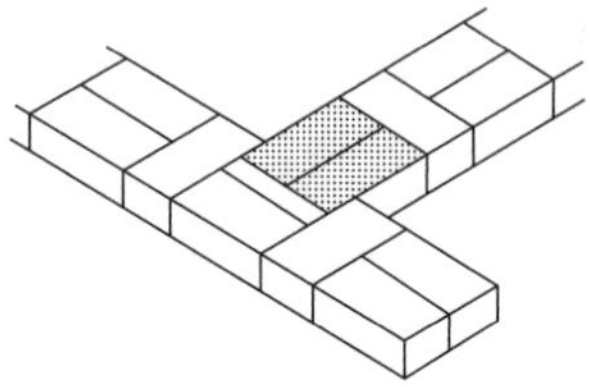
Rule 10 *In English bond, when a wall changes direction, the face bond changes from headers to stretchers or vice-versa in the same course. (See Fig. 4.14.)* *Exception: This rule will not apply when (a) walls of different thickness intersect, and (b) walling curves on plan, (c) where the change of direction is 112.5 mm break or less. (See Fig. 4.16.)*	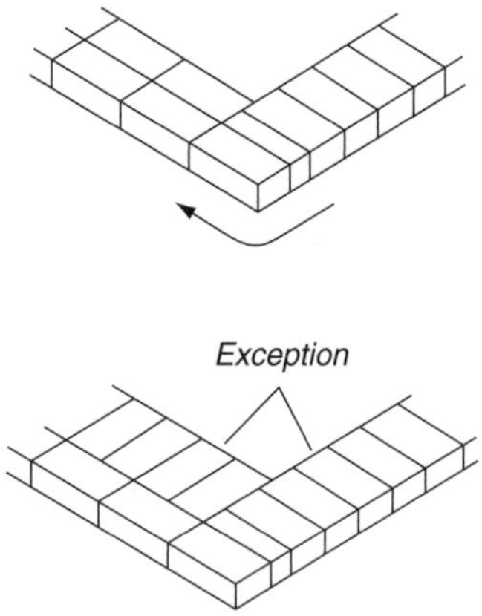
Rule 11 *The middle of thicker brick walls must be filled in with headers – to satisfy Rule 2. (See Figs 4.13 and 4.23.)* *Exception: Half bricks are used in 1½ brick thick Flemish bond walling.*	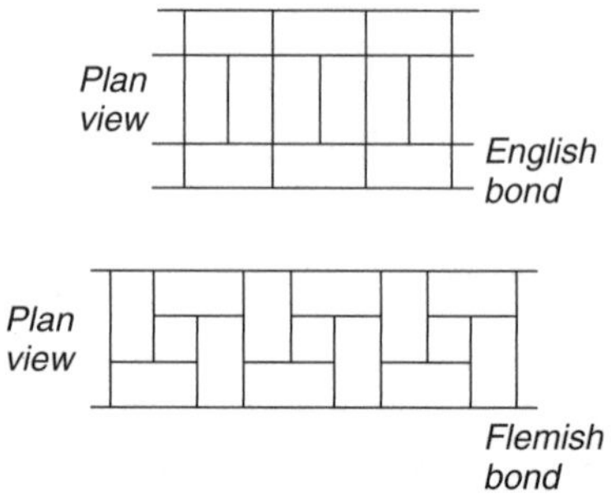

Figure 4.10 *(Continued)*

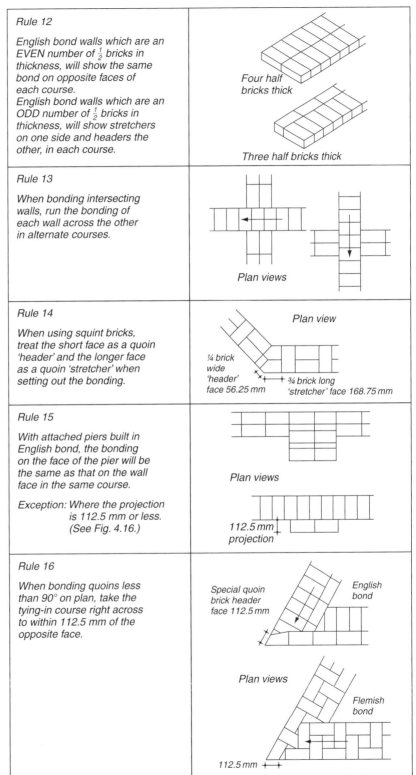

Rule 12

English bond walls which are an EVEN number of $\frac{1}{2}$ bricks in thickness, will show the same bond on opposite faces of each course.
English bond walls which are an ODD number of $\frac{1}{2}$ bricks in thickness, will show stretchers on one side and headers the other, in each course.

Four half bricks thick

Three half bricks thick

Rule 13

When bonding intersecting walls, run the bonding of each wall across the other in alternate courses.

Plan views

Rule 14

When using squint bricks, treat the short face as a quoin 'header' and the longer face as a quoin 'stretcher' when setting out the bonding.

Plan view

¼ brick wide 'header' face 56.25 mm

¾ brick long 'stretcher' face 168.75 mm

Rule 15

With attached piers built in English bond, the bonding on the face of the pier will be the same as that on the wall face in the same course.

Exception: Where the projection is 112.5 mm or less. (See Fig. 4.16.)

Plan views

112.5 mm projection

Rule 16

When bonding quoins less than 90° on plan, take the tying-in course right across to within 112.5 mm of the opposite face.

Special quoin brick header face 112.5 mm

English bond

Plan views

Flemish bond

112.5 mm

Figure 4.10 *(Continued)*

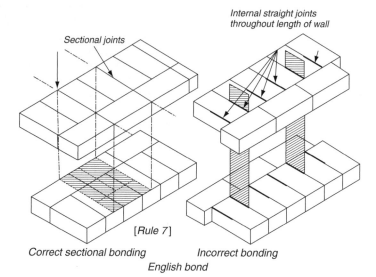

Internal straight joints
throughout length of wall

Sectional joints

[Rule 7]

Correct sectional bonding   Incorrect bonding

English bond

Figure 4.11

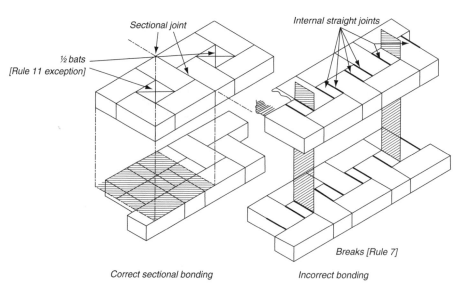

Sectional joint

Internal straight joints

½ bats
[Rule 11 exception]

Correct sectional bonding   Incorrect bonding

Breaks [Rule 7]

Flemish bond

Figure 4.12

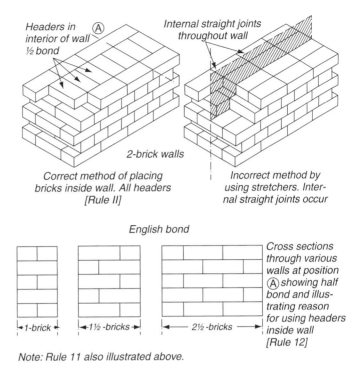

Headers in interior of wall ½ bond Ⓐ

Internal straight joints throughout wall

2-brick walls

Correct method of placing bricks inside wall. All headers [Rule II]

Incorrect method by using stretchers. Internal straight joints occur

English bond

Cross sections through various walls at position Ⓐ showing half bond and illustrating reason for using headers inside wall [Rule 12]

←1-brick→ ←1½ -bricks→ ← 2½ -bricks →

Note: Rule 11 also illustrated above.

Figure 4.13

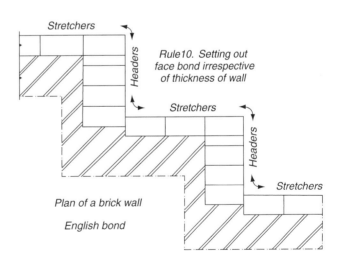

Stretchers

Headers

Rule10. Setting out face bond irrespective of thickness of wall

Stretchers

Headers

Stretchers

Plan of a brick wall

English bond

Figure 4.14

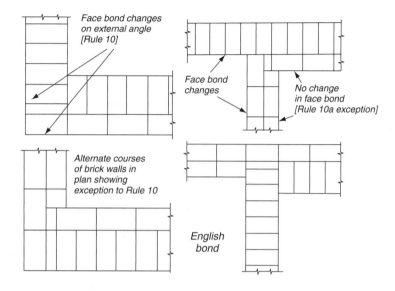

Face bond changes
on external angle
[Rule 10]

Face bond
changes

No change
in face bond
[Rule 10a exception]

Alternate courses
of brick walls in
plan showing
exception to Rule 10

English
bond

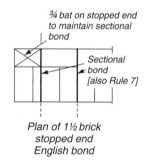

¾ bat on stopped end
to maintain sectional
bond

Sectional
bond
[also Rule 7]

Plan of 1½ brick
stopped end
English bond

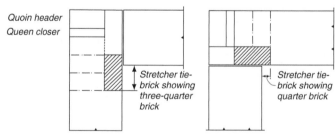

Quoin header
Queen closer

Stretcher tie-
brick showing
three-quarter
brick

Stretcher tie-
brick showing
quarter brick

Alternate courses of 1½ brick return angle in plan

Figure 4.15

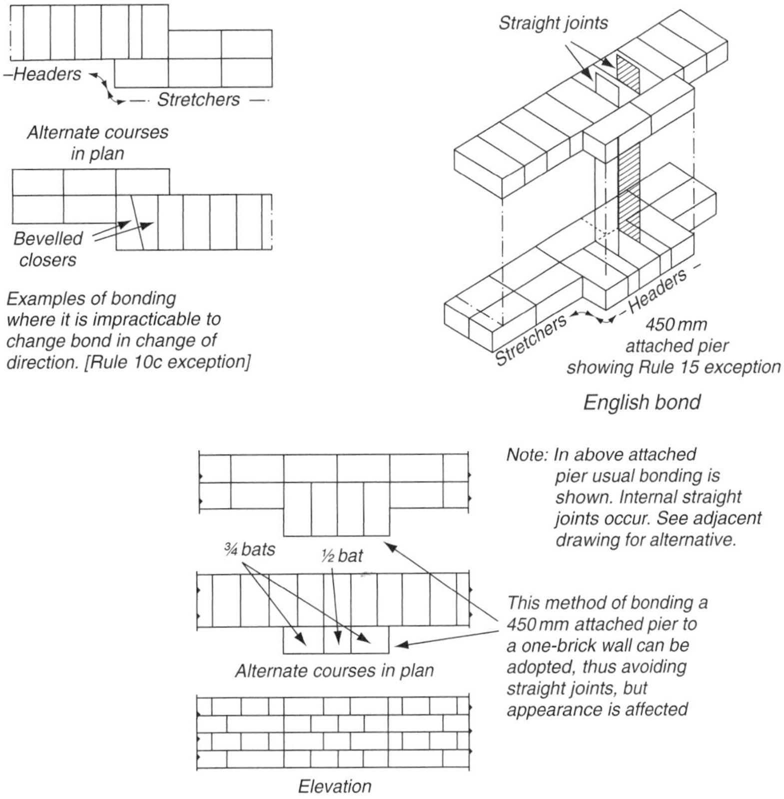

Headers
Stretchers
Alternate courses in plan
Bevelled closers
Examples of bonding where it is impracticable to change bond in change of direction. [Rule 10c exception]
Straight joints
Stretchers
Headers
450 mm attached pier showing Rule 15 exception
English bond
Note: In above attached pier usual bonding is shown. Internal straight joints occur. See adjacent drawing for alternative.
¾ bats
½ bat
This method of bonding a 450 mm attached pier to a one-brick wall can be adopted, thus avoiding straight joints, but appearance is affected
Alternate courses in plan
Elevation

Figure 4.16

Rules of bonding

When working out bonding arrangements, the bricklayer makes use of a number of Rules of Bonding, which are used as a guide. Like all good rules, there are permitted exceptions.

Use whole bricks wherever possible; when the first two courses of Stretcher, English and Flemish bonds have been set out correctly they will repeat themselves and the perpends (vertical joints) in every other course will be upright or 'plumb'. The bricklayer checks this by plumbing perpends at every 900 mm or so along the wall, course by course; see Fig. 4.17.

Problem 1

Set out in plan the alternate courses of a 1½-brick return angle in English bond.

First draw the outline of the problem to be solved (Fig. 4.18), then set out the first course on the external or internal angle, whichever is selected as face side. In ½-brick and 1-brick walls only one face side is obtainable. Walls thicker than one brick have two face sides, the thickness of the wall joint being adjustable to allow for the variation in brick size. Setting out in this case begins on the more important side.

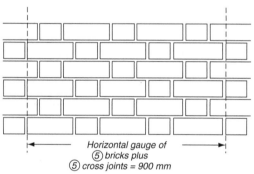

*Perpends plumbed with
pencil marks at 900 mm
intervals*

*Constant
vertical
gauge of
④ courses plus
④ bed joints
= 300 mm*

*Horizontal gauge of
④ stretchers plus
④ cross joints = 900 mm*

Stretcher bond

*Coordinating dimensions for brickwork which take into
account variations in size of bricks. The same principles
shown here on an elevation of stretcher bond apply
to other face bonds*

*Horizontal gauge of
⑤ bricks plus
⑤ cross joints = 900 mm*

Flemish bond

Figure 4.17 Plumbing
perpends

Beginning this problem with the external angle – Rules 4 and 10 apply
(Fig. 4.19). Continue by placing the tie-brick – Rule 7 (Fig. 4.20). Complete
'backing in' – Rule 7 – and, on stopped end use a ¾ bat (Fig. 4.21). If no
stopped end occurs or the wall is not of this particular length, i.e. 900 mm,
the tie-brick can be arranged as in Fig. 4.22.

Problem 2

Set out in plan the alternate courses of a two-brick return angle in English
bond (Fig. 4.23).

Outline the problem to be solved.

Set out the external face side, applying Rules 4 and 10. Continue by
setting tie-brick, and 'backing-in' – Rules 7, 10 and 12. Fill in the interior
of the wall – Rule 11.

Problem 3

Set out in plan the alternate courses of a 1½-brick return angle in Flemish
bond.

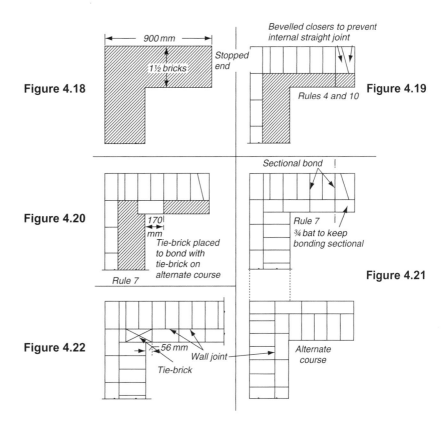

Figure 4.18

900 mm

1½ bricks

Stopped end

Figure 4.19

Bevelled closers to prevent internal straight joint

Rules 4 and 10

Figure 4.20

170 mm

Tie-brick placed to bond with tie-brick on alternate course

Rule 7

Figure 4.21

Sectional bond

Rule 7

¾ bat to keep bonding sectional

Figure 4.22

56 mm

Wall joint

Tie-brick

Alternate course

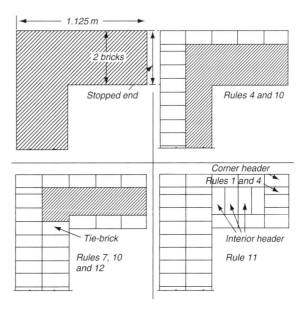

1.125 m

2 bricks

Stopped end

Rules 4 and 10

Tie-brick

Rules 7, 10 and 12

Corner header

Rules 1 and 4

Interior header

Rule 11

Figure 4.23

Outline the problem to be solved (Fig. 4.24).

Set out the external face side, applying Rules 4 and 7 (Fig. 4.25).

Place the tie-brick – Rule 7 – (Fig. 4.26), and complete problem.

Note the two headers together on the internal face side (Fig. 4.27); this is termed 'broken bond', and the position shown is the most usual and the best for this particular problem. The apprentice will find it is possible to place the broken bond nearer the angle and in some bonding problems this arrangement would probably be more suitable.

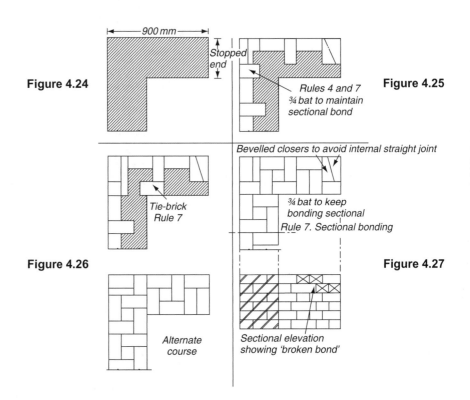

Figure 4.24

Figure 4.25

Figure 4.26

Figure 4.27

Understanding brick dimensions

The *co-ordinating size* of clay bricks, inclusive of mortar joints, is given in BS 3921:1985 as 225 mm long by 112.5 mm wide by 75 mm deep; see Fig. 2.4.

The *work size*, for which brick manufacturers aim (also given in the BS) is 10 mm less than each of those dimensions, to provide a 'joint allowance'. The work size for each standard metric brick is therefore 215 mm long by 102.5 mm wide by 65 mm deep.

The *actual sizes* of bricks as delivered vary slightly, and these variations must be absorbed by adjusting the thickness of mortar joints.

The individual differences in brick dimensions (larger or smaller than the work size) are mainly due to differing rates of drying and shrinkage of the naturally occurring clay or shale – the raw material. This is where the skill of the bricklayer is important. Opening or tightening up cross joints must be carried out carefully and evenly, so that:

(i) it is not noticeable in the finished wall;
(ii) plumb perpends are maintained at three or four brick intervals for the full height of the wall;
(iii) the co-ordinating size of the brickwork is maintained.

Figure 4.17 shows how varying sizes of brick must be kept within vertical and horizontal gauge.

Broken bond

This occurs where the length of a wall or pier, from end to end, does not work out to suit the bond pattern exactly.

In accordance with Rule 3, any broken bond should be located towards the middle of a wall or pier. Face work should be set out allowing a 225 mm space for each stretcher with its cross joint. Similarly, allowing 112.5 mm for each header plus joint.

On long walls there may be scope for slight adjustment of cross-joint thickness, from this basic module of 225 mm, in order to avoid broken bond and so achieve economy and a good appearance. (For example, tightening up 1 mm on every cross joint in the first course of a stretcher bond wall 8.95 m long would allow 40 whole bricks to be fitted into that length, using 9 mm rather than 10 mm joints.)

Broken bond can arise for two reasons.

(i) Where the architect's overall dimension of a wall does not equal a number of whole bricks. The cut portion of a brick located near the middle of a wall must never be smaller than a half bat, that is 102.5 mm; see Fig. 4.28.
(ii) Where the wall length does not need any cut bricks at the centre, but is still not a convenient length to maintain the correct repetition of bond pattern; see Fig. 4.29.

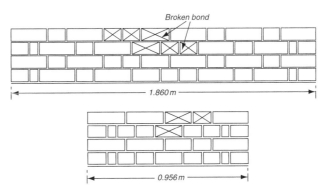

Figure 4.28 Location of broken bonding (inconvenient dimensions – solved using cut bricks)

For whichever reason broken bond applies, once located in the first two courses of a wall, the perpends of this broken bond must be plumbed carefully for the *full* height of the building elevation.

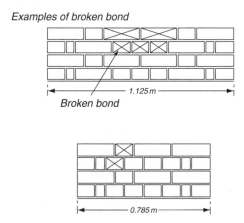

Examples of broken bond

Broken bond

— 1.125 m —

— 0.785 m —

Figure 4.29 Location of broken bonding (using whole bricks)

Setting-out facework in a wall without openings

Before setting-out the bond on the face side of any wall, it is wise to make a 'gauge rod'. This should be made of straight, smooth timber 50 mm × 50 mm in cross section, and approximately 3 m long. Along one face fine sawcuts carefully made at 225 mm intervals will allow stretcher bond to be set out consistently from end to end, leaving any broken bond near the middle of the wall.

On another surface of the rod, sawcuts at 75 mm intervals serve to check regular vertical gauge when bricklaying commences and quoins are raised. Use of this dual-purpose 'storey rod', for checking horizontal and vertical gauge, avoids potential risk of error when using a measuring tape.

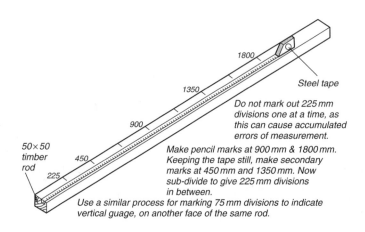

1800

Steel tape

1350

Do not mark out 225 mm divisions one at a time, as this can cause accumulated errors of measurement.

900

Make pencil marks at 900 mm & 1800 mm. Keeping the tape still, make secondary marks at 450 mm and 1350 mm. Now sub-divide to give 225 mm divisions in between.

50 × 50 timber rod

450

225

Use a similar process for marking 75 mm divisions to indicate vertical guage, on another face of the same rod.

Figure 4.30 Marking out a gauge or storey rod

Setting out a facework wall which includes openings

The principle here is to preplan window and door openings *before* a face brick is bedded. It is no good raising a facework wall to window sill level, then asking 'Right, now where do the windows go?'

If you do this, you will not get continuous perpends, straight and plumb, from top to bottom of the wall! First, mark out the location and widths of window and door openings by measurement on the ground beam or substructure brickwork; see Fig. 4.31.

'Reveal bricks' should be placed dry as indicated in Fig. 4.31. The bond pattern is then laid out to left and right of these fixed-point reveal bricks.

Keeping to the co-ordinating dimension of 225 mm horizontal gauge, any broken bond will be chased along to an approximately central point between the reveal bricks.

This system will locate any broken bond centrally under windows and/or in the middle of window piers.

When the perpends of each reveal brick are plumbed, as the wall is raised up, there will be a whole brick conveniently in place to form the vertical sides of every window opening when sill level is reached. Figure 4.32 shows some further variations when setting openings in facework.

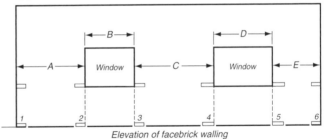

Providing architect's chosen dimensions A to E are multiples of 225 mm, then no broken bond will occur anywhere with stretcher and English bonding. Different dimensions for A to E will operate for Flemish or other bonding patterns, if broken bonding is to be avoided.

Elevation of facebrick walling
Reveal-brick stretchers 1 to 6 identified at ground level before facework commences. These become firm perpend plumb points for full height of walling.

Figure 4.31 Setting out facework

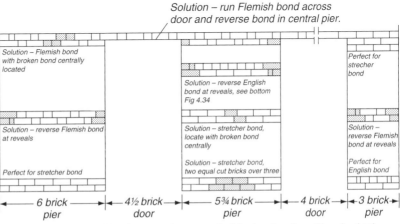

Figure 4.32 Setting out facework, various bonds. Further examples of using broken bond, or avoiding it with reverse bond

Once set out at ground level, perpends or reveals & broken bond must be plumbed for the full height of the walling.

Reducing the visual effects of broken bond

Broken bond in a facework wall can be visually disturbing above or below each window. The effect can be minimised if the mortar joint colour matches the brick colour. Light coloured mortar with dark bricks makes broken bond very obvious. This is not the bricklayer's fault, as broken bond is not incorporated by choice. Broken bonding costs time and money, both to set out and to cut the necessary bricks for the full height of the walling. If the widths of window and door openings are made to suit brick sizes *and* the chosen bond pattern, then there will be no broken bond at all. Figure 4.33 indicates a method of making broken bond less obvious in a stretcher bond wall.

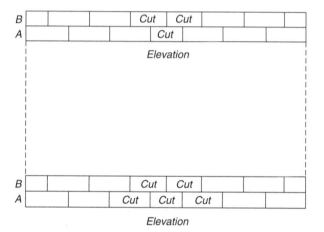

If the single cut brick in course 'A' of the upper panel is turned into three equal size cut bricks in course 'A' of the lower panel, then the broken bonding is much less obvious.

Figure 4.33 Broken bond in stretcher bond walling

Reverse bond

Occasionally it is possible, when setting out facework, to avoid broken bond at the centre of a wall or window pier by 'reversing' the bond. This means operating the exception to Rule 3. Experience tells the foreman bricklayer where and when it is possible to use this. Figure 4.34 gives three examples where reverse bond has been used as an alternative to broken bond.

It is not advisable to make use of reverse bond when setting out facework if a building has contrasting-coloured facing bricks ('dressings') at the sides of window and door openings. This would upset the balance of appearance; see Fig. 4.35.

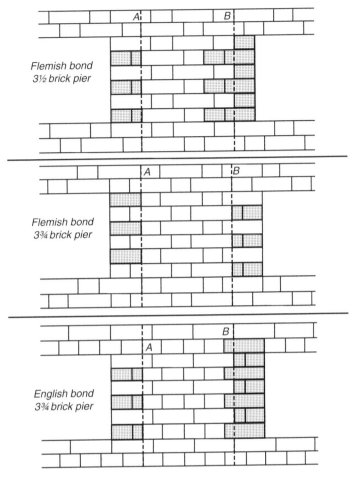

Notice how the firm perpends A and B (dotted) continue unbroken for the full height of the elevation. This fundamental principle is most important, particularly with light-colour mortar and dark-colour bricks and vice versa.

Note also how, in each case, the reverse bond (shaded) disappears above and below window openings, unlike setting out broken bond, which will be continuous from top to bottom of wall elevation.

Figure 4.34 Three examples of facework, where reverse bond has been used to avoid broken bond in window piers

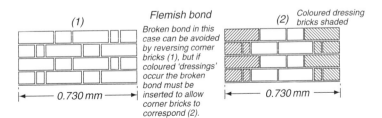

(1)

Flemish bond

Broken bond in this case can be avoided by reversing corner bricks (1), but if coloured 'dressings' occur the broken bond must be inserted to allow corner bricks to correspond (2).

(2) Coloured dressing bricks shaded

← 0.730 mm →   ← 0.730 mm →

Figure 4.35 Broken bond may be inevitable, particularly where contrast colour bricks are required at window reveals

Bonding examples

From the information already given in this chapter, the apprentice will readily understand the illustrated examples which follow. These will serve to introduce some craft terms, in addition to showing some of the problems which confront a bricklayer in practice.

An offset in the plan view of a wall may be referred to as a double return, because there are two quoins or returns formed, see Fig. 4.7. Alternatively, it may be called a 'break' in the line of the wall. See typical bonding arrangements in Fig. 4.36.

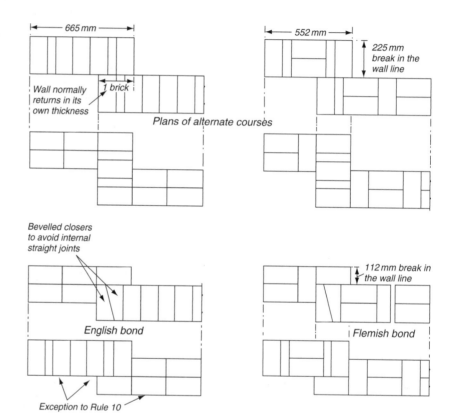

Figure 4.36 Bonding double returns, or breaks, in English and Flemish bond

Note the two methods of dealing with a ½ brick recess reveal in Flemish bond. There is less cutting in 'A' (Fig. 4.37) and it is probably the most practical. In any case, the cutting of reveal bricks is always difficult, especially in the case of hand-made and engineering bricks, the former creating excessive waste and the latter being almost impossible to cut with hammer and bolster.

The best method is to use purpose-made bricks (see Fig. 14.4, BD.1, BD.2 and BD.3). If these are not available, see Fig. 4.38 for hand cutting by hammer and bolster.

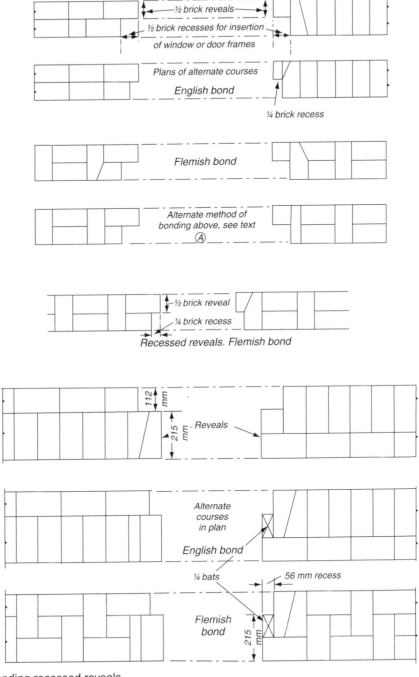

Figure 4.37 Bonding recessed reveals

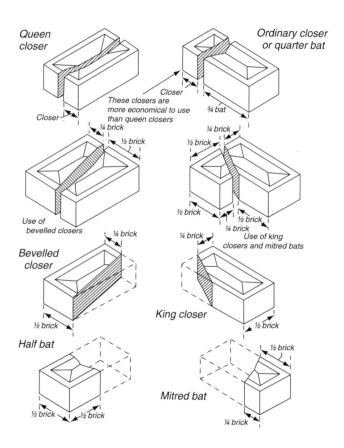

Figure 4.38 Definitions of cut bricks

Figure 4.39 shows two methods of dealing with the same problem. Number 2 is the better method, cut bricks being avoided on the face of the brick wall.

Figures 4.40 to 4.42 show attached piers. Note the arrangement of bricks in bonding a 330 mm attached pier in Flemish bond (Fig. 4.40, lower half). The appearance of the attached pier is English bond, nevertheless it is stronger and looks better than the method sometimes used (Fig. 4.41), which is often mistakenly described as Flemish bond, but which is, in fact, half-bonding.

Figure 4.42 shows two ways of arranging a 440 mm attached pier in Flemish bond. The top method uses a broken bond on the straight face; the other avoids this.

Figures 4.43 and 4.44 illustrate junctions in 1½-brick walling. Note the two arrangements of bricks in junctions.

Figures 4.45 to 4.47 show a brick wall introducing a 1½-brick double return angle and two methods of finishing a stopped end in a 1½-brick wall. It is usual, where stopped ends are to be rendered or partly hidden by a door or window frame, to build with the elevation showing half-bond. Less cutting is needed than in the case of the stopped end on buttress, where an elevation showing quarter-bond is necessary to maintain a correct appearance.

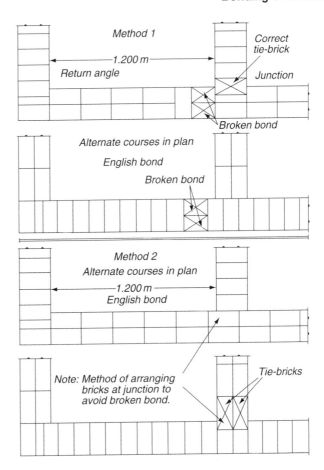

Figure 4.39 Economy in solving problems of bonding

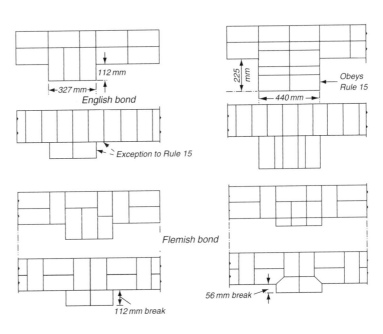

Figure 4.40 Attached piers

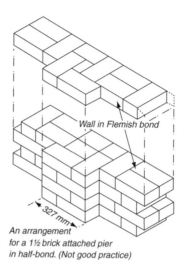

Figure 4.41 The stretcher bond on the face of the pier does not match the Flemish bond walling behind

Wall in Flemish bond

327 mm

An arrangement for a 1½ brick attached pier in half-bond. (Not good practice)

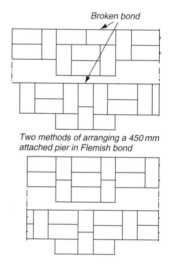

Broken bond

Two methods of arranging a 450 mm attached pier in Flemish bond

Figure 4.42

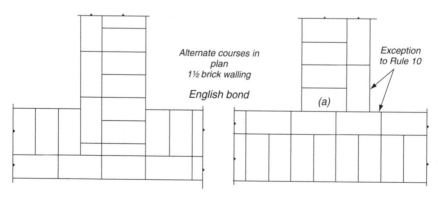

Alternate courses in plan
1½ brick walling

English bond

Exception to Rule 10

(a)

Figure 4.43 Junction walls

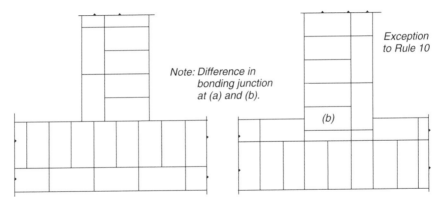

Note: Difference in
bonding junction
at (a) and (b).

Exception
to Rule 10

(b)

Figure 4.44 Junction walls – alternative bonding

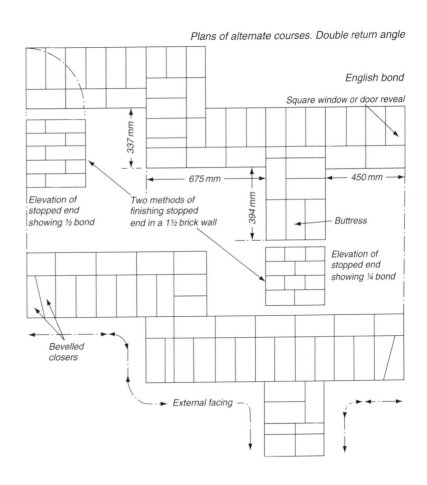

Plans of alternate courses. Double return angle

English bond

Square window or door reveal

337 mm

675 mm

450 mm

394 mm

Elevation of
stopped end
showing ½ bond

Two methods of
finishing stopped
end in a 1½ brick wall

Buttress

Elevation of
stopped end
showing ¼ bond

Bevelled
closers

External facing

Figure 4.45 Double returns
in 1½ brick walling

Figures 4.48 and 4.49 show two problems in which rules cannot be strictly adhered to. The figures are self-explanatory. Examine them carefully and set out the problem.

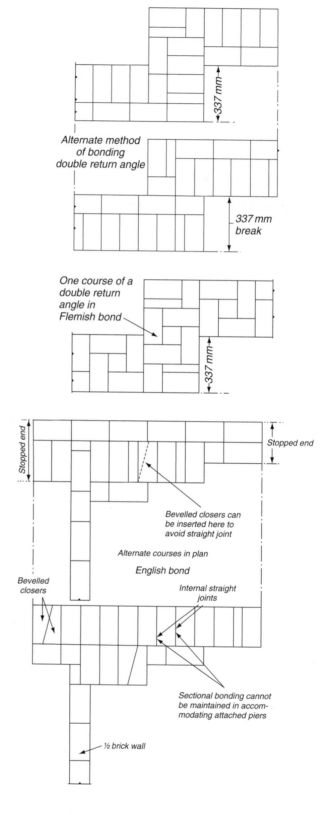

Figure 4.46 Alternative to Fig. 4.45

Alternate method of bonding double return angle

337 mm

337 mm break

Figure 4.47

One course of a double return angle in Flemish bond

337 mm

Stopped end

Stopped end

Bevelled closers can be inserted here to avoid straight joint

Alternate courses in plan

English bond

Bevelled closers

Internal straight joints

Sectional bonding cannot be maintained in accommodating attached piers

½ brick wall

Figure 4.48 An interesting example of bonding

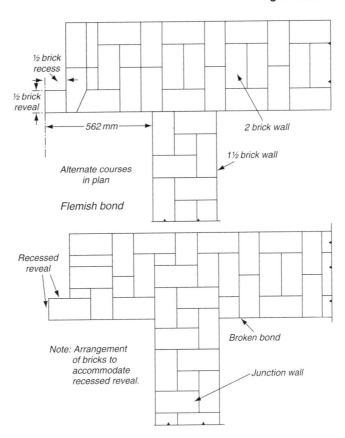

½ brick recess

½ brick reveal

562 mm

2 brick wall

1½ brick wall

Alternate courses in plan

Flemish bond

Recessed reveal

Broken bond

Note: Arrangement of bricks to accommodate recessed reveal.

Junction wall

Figure 4.49 Another interesting bonding example

Summary

This chapter has set out the basic Rules of Bonding applicable to solid walling, giving examples of walls up to two bricks in thickness. Although a great deal of brickwork is in the form of cavity wall construction involving a half brick thick, stretcher bonded outer leaf only, it is important for the bricklayer to be confident with the Rules of Bonding as applied to walls of greater thickness than 102 mm and to be familiar with other bond patterns.

The next chapter introduces other face bonds and applications of bonding details.

5   Bonding of blockwork

As mentioned previously blocks are produced in a range of shapes and sizes. The face side is usually 450 mm × 225 mm, the thickness varies from 37 mm up to 1225 mm and the weight from 6.3 to 15 kg.

They are produced in solid, hollow and multicut format. Multicuts enable a bolster cut to be made without wastage.

Types and method of construction

Blockwork has become very popular not only as a quick method of producing inside walls and partitions but as a facing material in its own right.

Blocks are very economic in both handling and purchasing compared with bricks. Apart from their size the craft operations in erecting walls and corners are as for brick.

However, there are several ways in which blockwork can be erected satisfactorily, depending upon the thickness of the blockwork:

1. When internal walls have been set out fix profiles at angles and indents
2. To prevent buckling limit the height of wall to six courses per day
3. Support long walls to prevent buckling.

Setting out

It is important to dry bond the first course to avoid awkward cuts as shown in Fig. 5.1. Set out to a predetermined chalk line.

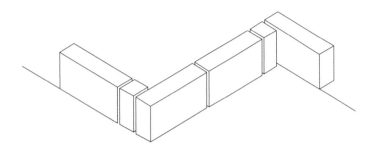

Figure 5.1 Dry bonding first course

Many types of blocks are difficult to cut, although many specials are available such as quarter, half, three-quarter and reveal/corner blocks; see Fig. 5.2. Refer to Chapter 2 Materials.

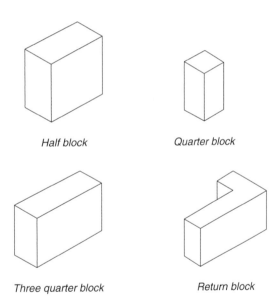

Half block   *Quarter block*

Three quarter block   *Return block*

Figure 5.2 Types of specials

Remember

Never mix different block types in the same wall and never use bricks as they will cause a reduction in the insulation value of the cavity wall and could cause problems with drying shrinkage and pattern staining due to different absorption of various materials.

As mentioned in Chapter 3 Tools, a masonry saw is available to assist in cutting blocks although many sites now have masonry bench saws.

It is an offence under the Health and Safety at Work Act to use a masonry bench saw if you are under the age of 18. Even when you are 18 and over the Act states that you must have been instructed in its safe use and carry an Abrasive Wheels Certificate.

Bonding

The bond used for blockwork should be half bond where possible; see Fig. 5.3. On no account should the bond be less than a quarter block length.

Work to tight lines and check for plumb regularly. The maximum lift of blocks completed in one day should not exceed 1.5 m, which is normally six courses.

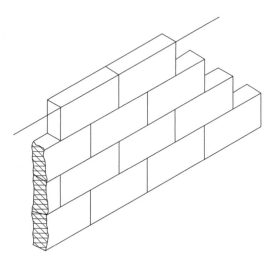

Figure 5.3 Half bonding blockwork

Quoins

Quoins are erected and racked back as for brickwork and a quarter block is required next to the corner block to obtain half bond; see Fig. 5.4.

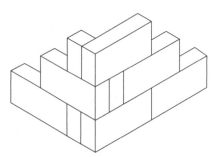

Figure 5.4 Quoin with quarter blocks

Another method is to cut the first block to a three-quarter to obtain the bond; see Fig. 5.5.

Special corner blocks could be used but are an extra expense; see Fig. 5.6.

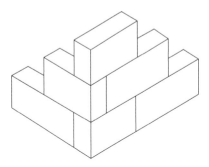

Figure 5.5 Quoin with three-quarter blocks

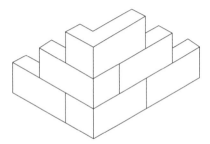

Figure 5.6 Quoin with return blocks

Plumbing blockwork

Plumbing blockwork is more difficult than plumbing brickwork as the wall rises faster and the mortar tends to squeeze out the soft mortar bed joints causing the wall to move out of plumb.

Always ensure blocks are fully bedded with nominal 10 mm bed and cross joints.

On no account tap the blocks sideways – this will only result in the bed joint opening up on one side. Always plumb the wall by tapping down on the high side of the block; see Fig. 5.7.

Broken bonds

Broken bonds may have to be adopted when wall lengths are not equal to block sizes. It will mostly be the case that block walls will require some method of broken bond. It is important never to use cuts less than half a block as this can create a weak plane in the wall and if drying shrinkage occurs it may result in cracks appearing in the wall as shown in Fig. 5.8.

Blocks can be cut with a club hammer and bolster chisel, a masonry saw or a mechanical bench saw. Try to make use of any special blocks

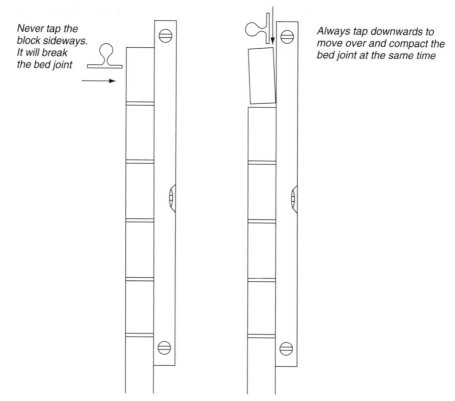

Never tap the block sideways. It will break the bed joint

Always tap downwards to move over and compact the bed joint at the same time

Figure 5.7 Plumbing blockwork

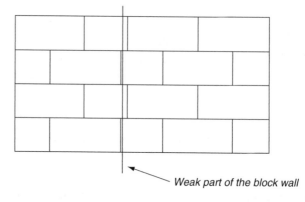

Weak part of the block wall

Figure 5.8 Broken bond

that are available before resulting in cutting. Always try to achieve a neat cut which retains the 10 mm cross joint. Overlarge joints can again result in cracks appearing in the wall due to drying shrinkage.

Broken bonds of less than half lap can be avoided using blocks, as in Fig. 5.9.

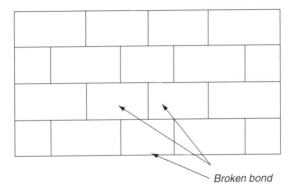

Figure 5.9 Broken bond

Reverse bond

This is when the blocks at each end of the wall are different.

The following front elevation, Fig. 5.10, shows a reverse bond being used.

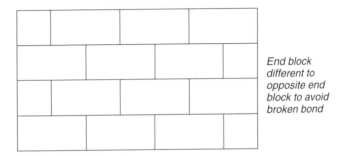

End block different to opposite end block to avoid broken bond

Figure 5.10 Reverse bond

When cut blocks are being used it is more economical to use three-quarter, half and quarter ones provided by the manufacturer.

Blocks can be cut with a masonry saw – when using lightweight blocks – or lump hammer and bolster when using dense blocks.

Whenever cut blocks are required they should always be of the same material.

Junction walls

It is often inconvenient to build junction walls at the same time as the main wall so there will be the need during construction to make provision for work at a later date.

Indents should be left in the main wall to receive the junction wall later. These indents should be marked out correctly and kept plumb throughout the building of the block wall. The minimum lap at 'T' junctions should be quarter lap. An allowance of 20 mm is usually made over the width of the block to allow the blocks on the junction wall easy access into the indent; see Fig. 5.11.

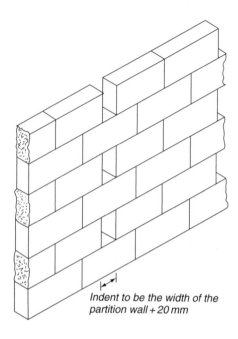

Indent to be the width of the partition wall + 20 mm

Figure 5.11 Indents

Other methods of providing a tie to junction walls which is acceptable is metal reinforcement; see Fig. 5.12. This is built into the main wall on every alternate bed joint. This method could help avoid numerous cutting at a junction of two block walls.

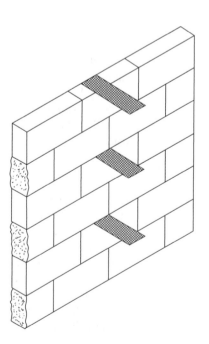

Figure 5.12 Reinforcing mesh

This method is recommended for bonding block walls together when both walls are constructed of different block types especially when the blocks have different shrinkage properties.

Bonding to brick walls

There will be the need to bond blocks to brick walls. This can be done by leaving indents in the brickwork or by fixing a proprietary wall connector as shown in Fig. 5.13.

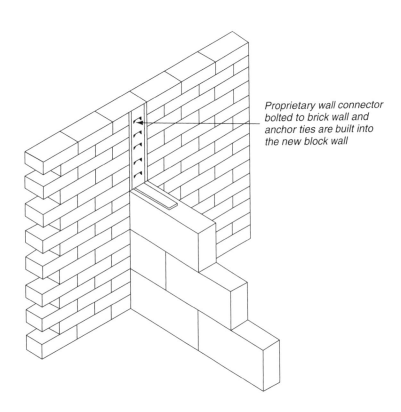

Proprietary wall connector bolted to brick wall and anchor ties are built into the new block wall

Figure 5.13 Proprietary wall connectors

Pillars and piers

It may be necessary to support long walls by having attached piers designed into them. There are several methods of bonding attached piers.

The use of blocks laid flat could be adopted but this affects the bond on the face of the wall; see Fig. 5.14.

A better method is to have a pier the same size as the blocks, to avoid cutting, and lay three together; see Fig. 5.15.

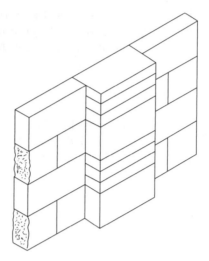

Figure 5.14 Attached pier

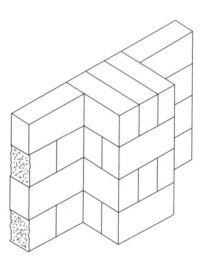

Figure 5.15 Attached pier with 150 mm blocks

Thin joint masonry

Continual research has enabled leading block manufacturers to develop a thin joint mortar which in conjunction with a larger block size enables internal walls to be constructed quicker.

This system is an alternative to using 10 mm sand and cement mortar bed and cross joints. The thin joint mortar is supplied as a dry pre-mixed powder in 25 kg bags and is applied with a special spreader, 3 mm thick. It begins to set within 10 minutes and approaches full strength in just 1 to 2 hours. The thin joint mortar is mixed with water as required.

The approach to using the thin joint system is different to the sand and cement system. The blocks have to be produced to a tolerance to allow them to be used in thin joint systems and all cut blocks need to be cut to the exact size allowing only 3 mm for the joints. A masonry saw or mechanical saw could be used, not a club hammer and bolster chisel as this would lead to greater joint widths than 3 mm.

Stability of the walls is greater than with sand and cement so therefore greater height can be reached with very little or no support required.

The first course is usually laid onto the dpc using traditional sand and cement mortar to achieve a level datum to work to; see Fig. 5.17. This is the most important course as it is impossible to gain or reduce height with thin joint mortar as easy as you can with traditional mortar. This course should be allowed to set overnight and be ground level before commencing with the thin joint system.

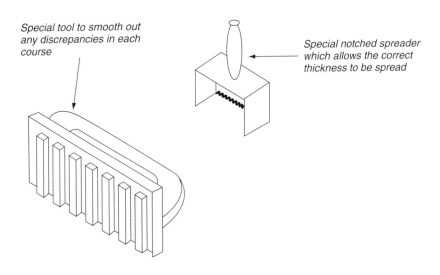

Special tool to smooth out any discrepancies in each course

Special notched spreader which allows the correct thickness to be spread

Figure 5.16 Special tools

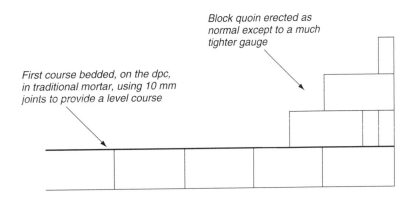

Block quoin erected as normal except to a much tighter gauge

First course bedded, on the dpc, in traditional mortar, using 10 mm joints to provide a level course

Figure 5.17 First course and thin joint quoin

The skills required to build with the thin joint system are very similar but it is necessary to maintain regular checks for level, plumb and line; see Fig. 5.18.

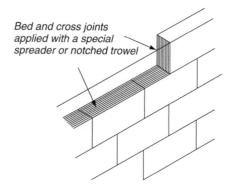

Bed and cross joints applied with a special spreader or notched trowel

Figure 5.18 Laying to line

The dry pre-mixed mortar should be mixed in a clean bucket to the rate of 25 kg of powder to 5.75 litres of clean water.

Special electric mixing tools are available to ensure correct consistency of the mortar. As with traditional mortar, once mixed, unused mortar should not be re-tempered or admixtures applied.

If mixed correctly thin joint mortar should remain workable for up to 4 hours. Ensure overmixing does not take place to prevent wastage.

The procedure for laying is the same as for traditional blockwork. The first course must be set out dry to ensure correct bond with minimum cutting.

The corners should be erected first and the walls run in as normal.

The thin joint mortar is spread using a special notched trowel or scoop to the correct thickness of 3 mm. The mortar bed should remain workable for 6 to 9 minutes and set within 10 minutes. Any minor adjustments should be made as soon as possible. Cross joints can be applied with the same equipment or the blocks could be dipped into the mortar to apply the cross joint.

One of the main differences is the height the wall can be built to. If the wall is the inner leaf of a cavity wall, normal wall ties cannot be used. The inner leaf can be built to wall plate level and the inner work can continue while the outer leaf of the cavity wall is being built.

Wall ties are placed at the normal positions but due to the bed joint not coinciding, special helical wall ties, shown in Fig. 5.19, are driven into the face of the blocks.

Insulation can be placed against the inner blockwork and the wall ties driven through the insulation. If partial insulation is used, special clips are inserted as normal. It is important to ensure the block wall has set and is stable before driving in wall ties.

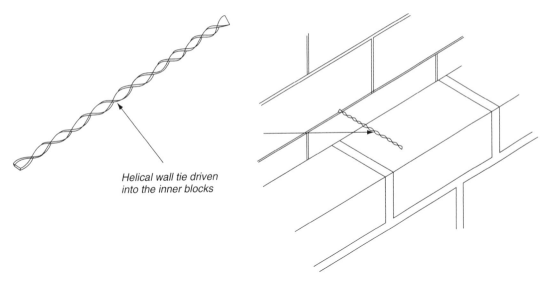

*Helical wall tie driven
into the inner blocks*

Figure 5.19 Helical wall tie

The benefits of the thin wall system can be greatly enhanced if jumbo block units are used. Jumbo block units are available 440 mm long and 430 mm high. These sizes can change between manufacturers.

6 Bonding details

This chapter deals with bonding details involving the use of bricks of special shape called plinths, and of squints, and where walls do not intersect at right angles, e.g. obtuse angle quoins of more than 90°; and of acute angles of less than 90° in plan. (Chapter 14 describes the range of BS special shape bricks available.)

The arrangement of face bonds other than English, Flemish and stretcher bonds is illustrated, followed by coverage of isolated piers, bond patterns for decorative panels, and the formation of brickwork curved on plan.

Plinths

Figures 6.1(a) to (c) show bonding details where substructure brickwork is set back, or 'set in' (in this case by 28 mm), for architectural effect, at the commencement of the superstructure, giving a building the appearance of a 'plinth' or base. In some buildings this 'plinth thickening' may be constructed from common bricks and rendered with cement and sand mortar afterwards.

Purpose-made plinth bricks, with a 45° sloping surface, are used to achieve this set-on of face brickwork for a few courses above ground level, by 56 mm per course. These plinth headers and stretchers are based upon the standard $215 \times 102 \times 65$ mm size to fit in with bonding and gauge of main walling. An 'external return' special plinth is shown in Fig. 6.1(a).

Plinths are always a source of anxiety to the foreman bricklayer, as difficulties arise in maintaining the correct bond on the face and at the same time avoiding unnecessary cuts and straight joints. Figure 6.1(a) shows a 28 mm set-on acting as a plinth, and Fig. 6.1(c) is the arrangement of

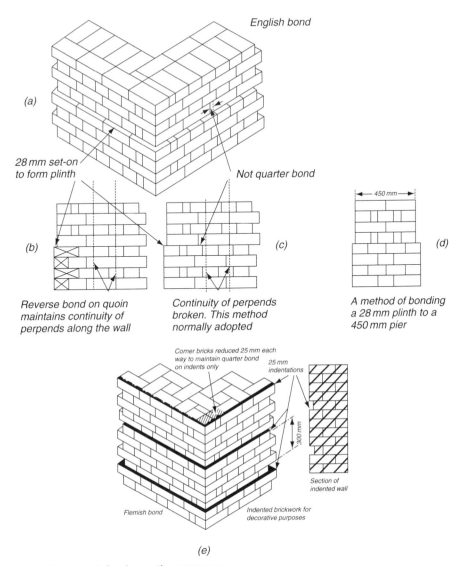

Figure 6.1 Indented brickwork for decorative purposes

bricks at the quoin. This is the method usually adopted. An alternative is shown (Fig. 6.1(b)) where continuity of perpends is an absolute necessity. Figure 6.2(a) should be carefully noted.

Preplan the bonding

The bonding arrangement for any plinth brickwork must be preplanned well in advance of starting work. A little time spent with squared paper, drawing it out beforehand, will avoid the embarrassment of straight joints later on.

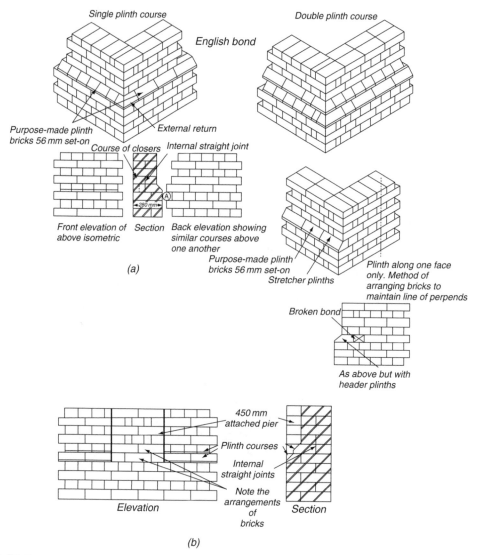

Single plinth course

English bond

Double plinth course

Purpose-made plinth bricks 56 mm set-on

External return

Course of closers   Internal straight joint

Front elevation of   Section   Back elevation showing
above isometric   similar courses above
                                one another

Purpose-made plinth
bricks 56 mm set-on
Stretcher plinths

Plinth along one face
only. Method of
arranging bricks to
maintain line of perpends

Broken bond

As above but with
header plinths

(a)

450 mm
attached pier

Plinth courses

Internal
straight joints

Note the
arrangements
of
bricks

Elevation   Section

(b)

Figure 6.2 Plinths

Draw in the face bonding required on squared paper in the numerical order shown in Fig. 6.3. Start above the plinth courses on both faces of a quoin.

Plumb bonding perpends down through the plinth courses. Adjust the face bonding from (3) towards (4) in Fig. 6.3, to avoid straight joints on both faces of the quoin, using broken bond and bevelled closers.

Figure 6.3, if carefully followed, should enable an apprentice to set out most plinth problems. Figure 6.2(b) shows the setting out of an attached

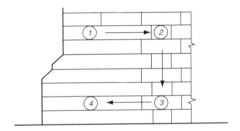

Figure 6.3 Preplanning bonding procedure for plinth brickwork

pier above plinth courses. The arrangement of bricks is sound and avoids unnecessary cutting.

Plinth bricks may also be used upside down to construct the brick corbel courses shown in Fig. 8.57, to give a more decorative effect.

Indented brickwork

Indenting one course along a whole elevation, as shown in Fig. 6.1(e), produces interesting shadow effects. Alternatively, this indenting can be restricted to approximately $1\frac{1}{2}$ bricks only in each direction at quoins, for the same purpose.

Decorative brick quoins

Figure 6.4 shows other ways of making the angles of a building look more obvious; this is sometimes described as 'rusticated' blocked corners and dressed corners.

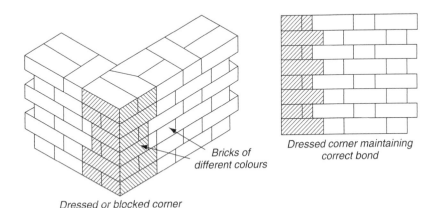

Bricks of different colours

Dressed corner maintaining correct bond

Dressed or blocked corner

Figure 6.4 Contrast colour bricks used at quoins

Acute angles (Fig. 6.5)

Note the method of planning quoin brick to obtain quarter-bond along wall. Owing to its small size, a closer is omitted from this corner; it would be difficult to cut and awkward when laying.

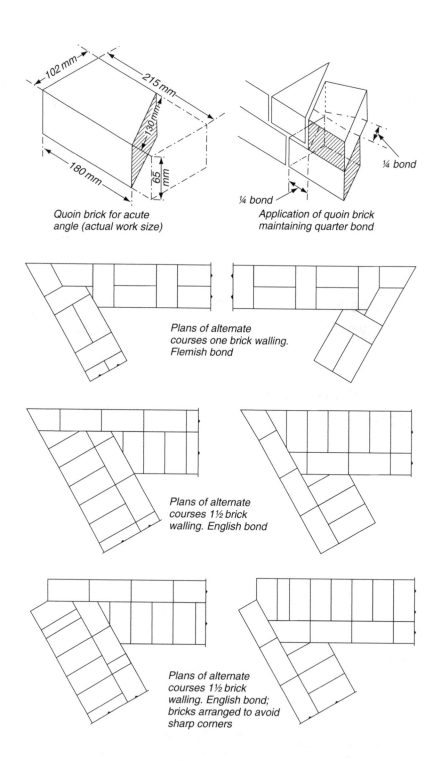

Quoin brick for acute
angle (actual work size)

Application of quoin brick
maintaining quarter bond

¼ bond

¼ bond

Plans of alternate
courses one brick walling.
Flemish bond

Plans of alternate
courses 1½ brick
walling. English bond

Plans of alternate
courses 1½ brick
walling. English bond;
bricks arranged to avoid
sharp corners

Figure 6.5 Acute angles

Obtuse angles or squint corners (Fig. 6.6)

Note the planning of quoin brick. Both acute and obtuse angle bricks are planned to conform to standard brick size and are most generally used. Others used are normally oversize and purpose made to the architect's specification. Note the alternate courses of brickwork at 'birdsmouth' angles, 'A' in the bottom element of Fig. 6.6. Careful craftsmanship is needed here to avoid the appearance of a straight joint, by emphasising the tie bricks to left and right when jointing up.

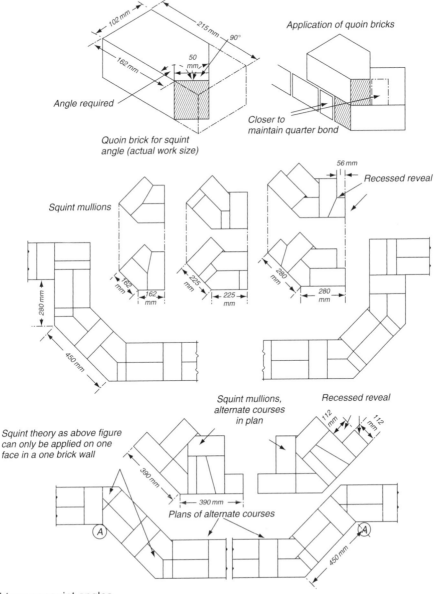

Figure 6.6 Obtuse or squint angles

Panel walls and isolated piers (Figs 6.7 and 6.8)

The panel walls are planned to show a broken bond. Note the latter's position in this case. The 330 mm isolated pier 'A' avoids straight joints but requires an excessive number of cut bricks. Pier 'B' has internal straight joints and shows half-bonding on two sides but is considered sound enough for most purposes and is therefore more generally adopted.

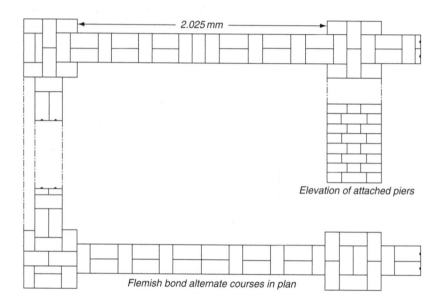

Figure 6.7 Panel walls

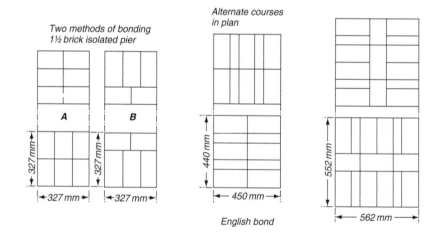

Figure 6.8 Isolated piers

Other types of bond

Single Flemish bond (Fig. 6.9)

This is a combination of English and Flemish bonds. One side of the wall shows Flemish bond and the other English, which being stronger is given preference in the thickness of the wall. Flemish bond is therefore only a facing and is bonded in by its headers on every other course.

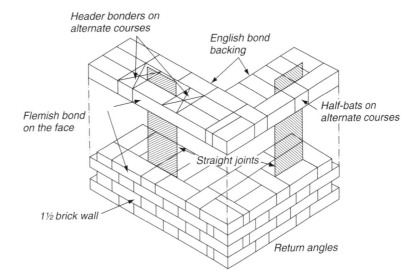

Figure 6.9 Single Flemish bond

Single Flemish bond is used:

1. for its combination of strength and appearance;
2. for economy of facing bricks, 'snap' headers being used on every other course;
3. if the architect wishes to show a different bond on each side of the wall to conform with the type of brick in use.

Dutch bond (Figs 6.10 and 6.11)

Dutch bond is similar to English bond in appearance, the difference being in the external corner. A three-quarter bat takes the place of the header and closer, and a half-bat or header is introduced into every other stretcher course next to the corner three-quarter.

Figure 6.10 Dutch bond

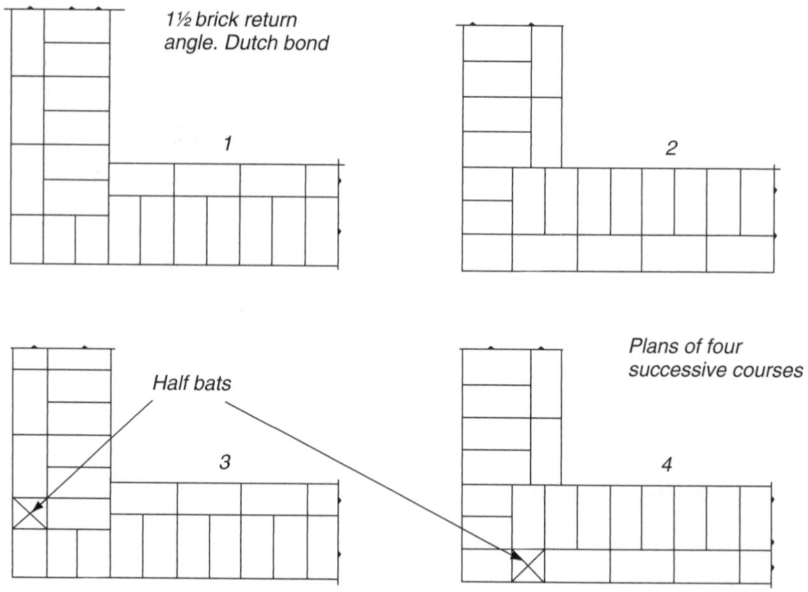

1½ brick return angle. Dutch bond

1

2

Half bats

3

4

Plans of four successive courses

Figure 6.11 Dutch bond example applied to 1½ brick walling

The advantages of this bond are:

1. bricks of a different colour from those being used for the main brick-work can easily be introduced into the wall face, forming patterns termed 'diapers'; see Figs 6.10 and 8.66;
2. loads are more evenly distributed owing to the fact that regular quarter brick offsets occur in the bond;
3. when using a multi-coloured brick, an attractive wave effect is obtained.

English cross bond

Similar to Dutch bond, except that a header and closer are introduced in place of a three-quarter corner brick (Fig. 6.12).

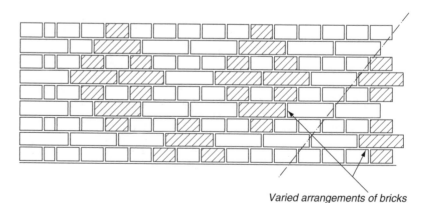

Figure 6.12 English cross bond

Varied arrangements of bricks

Monk bond

Two stretchers to one header in the same course. This facing bond introduces a zig-zag effect into the wall. Note that the bricks forming the pattern are exactly 1½ bricks apart as shown in Fig. 6.13. If this is specially noted, no difficulty will be experienced in maintaining the correct bond. For variations of monk bond or of '2 and 1' bond, see Fig. 6.14.

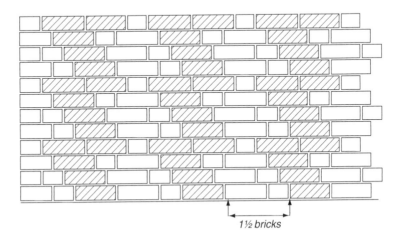

1½ bricks

Figure 6.13 Monk bond

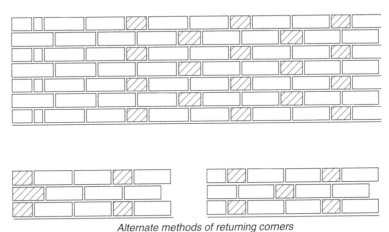

Alternate methods of returning corners

Figure 6.14 Variations of monk bond

Garden wall bonds

Facing bonds in which many more stretchers than headers are used. Several internal straight joints occur, but the advantages are:

1. fair faces can be obtained on both sides of a one-brick wall;
2. economises in facing bricks.

English garden wall bond

Consists of three courses of stretchers to one of headers (Fig. 6.15). An internal straight joint occurs throughout the length of the wall, three courses in height.

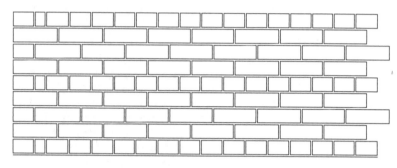

Figure 6.15 English garden wall bond

Note the half bonding of the stretcher courses

Flemish garden wall bond

Consists of three stretchers to one header in the same course. To maintain the correct bond, the header in one course must be in the centre of the middle stretchers in courses above and below (Fig. 6.16). In contrast to normal English and Flemish bonds this bond is stronger than English garden wall bond, the header bonders being more evenly distributed.

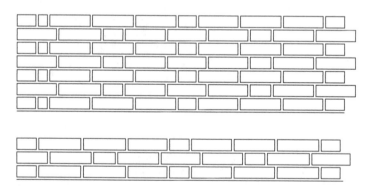

Figure 6.16 Flemish garden wall bond

Rat-trap bond

Bricks laid on edge in form of Flemish bond. Used for garden walls or where walls of a building are to be vertically tiled, or as it is termed 'tile-hung'. The pockets formed inside the wall can be left hollow or made solid according to instructions (Fig. 6.17).

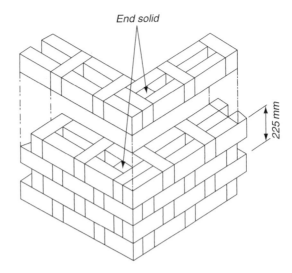

Figure 6.17 Rat-trap bond

Quetta bond

Used in walls $1^1/_2$ bricks or more in thickness, which are reinforced vertically with concrete and mild steel rods. The face bond is in either Flemish
or Flemish garden wall bond. It will be noted that small reinforced
concrete columns are formed at intervals along the wall (Fig. 6.18).

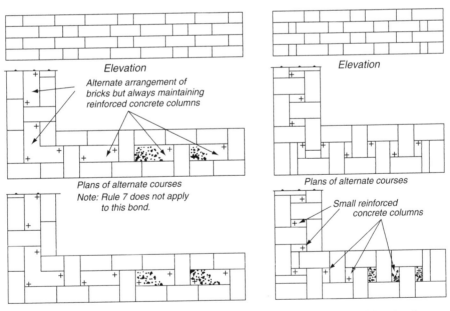

Figure 6.18 Quetta bond

Single herring-bone, double herring-bone, basket-weave and diagonal basket-weave bonds (Fig. 6.19)

Decorative bonds used in panel work for floors and paths or vertically in walls. Note the method of setting out the herring-bone bonds. This ensures cut bricks of the same size on all sides of the panel. For a small square or rectangular panel the setting out must commence from the centre of the panel.

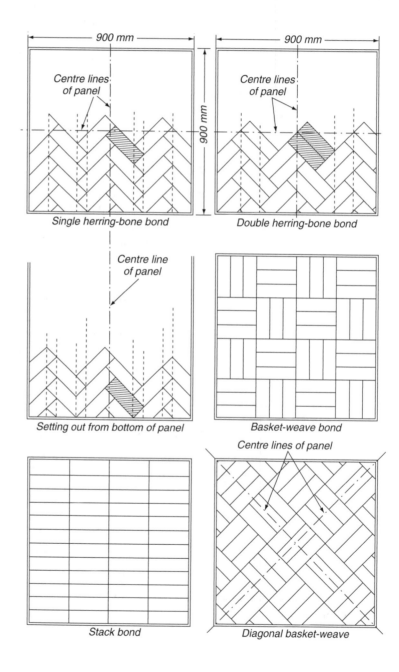

Figure 6.19 Decorative panels of brickwork

For a long narrow panel, the setting out can begin from the bottom centre. Many architects design decorative panels by combining these bonds, by introducing 'creasing' tiles, or by other arrangements of bricks.

Header bond (Fig. 6.20)

Used in the construction of circular brickwork. In the building of circular cesspools and similar work where a small radius occurs, purpose-made radiating bricks are preferable (see Fig. 14.4, RD.1 and RD.2), but in general construction the ordinary brick can be used. To avoid excessive cutting, the bond is arranged as shown in the figure, i.e. two half-bats and a one-brick bonder placed alternately. If uncut bricks are used, unduly wide cross joints would appear on one side of the wall (see the bottom right-hand element in Fig. 6.20).

With the use of either commons or hand-made facing bricks, a header bond is suitable, up to a radius of approximately 2250 mm. Beyond this measurement, the normal type of bond can be introduced. Where a glazed

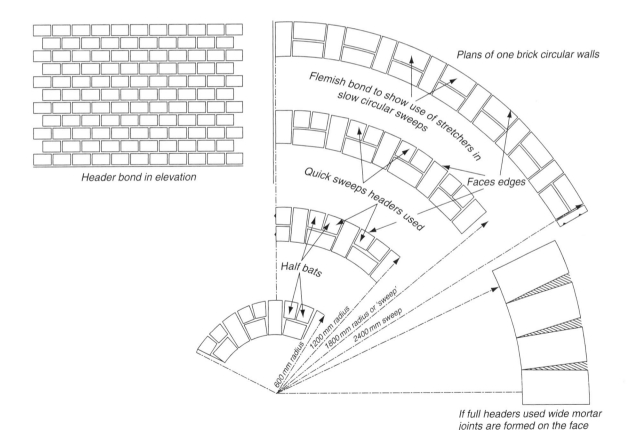

Header bond in elevation

Plans of one brick circular walls

Flemish bond to show use of stretchers in slow circular sweeps

Quick sweeps headers used

Faces edges

Half bats

600 mm radius

1200 mm radius

1800 mm radius or 'sweep'

2400 mm sweep

If full headers used wide mortar joints are formed on the face

Figure 6.20 Header bond

or similar brick is being used, it is necessary to extend the radius to approximately 4000 mm before adopting a normal bond. The brick, being regular in shape and size, tends to 'hatch and grin' (see Fig. 8.3(a)) if stretchers are used on a smaller radius.

The apprentice will encounter other kinds of bonds, but these are merely variations of those already mentioned. They have no specific name but are used by the architect when considered more suitable to use with the type of brick that has been specified.

7 Foundations

The foundations of a building are that part which is in direct contact with the ground.

The current Building Regulations require that the foundations of a building shall safely sustain and transmit to the ground the combined dead and superimposed loads in such a way as not to impair the stability or cause damage to any part of the building.

The ground or subsoil on which a building rests is called the 'natural foundation' or 'sub-foundation', and has a definite load-bearing capacity, according to the nature of the soil; see Fig. 7.1.

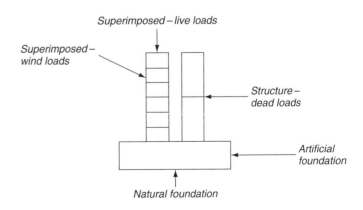

Figure 7.1 Loads

Table 7.1 Types of subsoils

Type of subsoil	Condition of subsoil
Class 1. Rock	Sandstone, limestone or firm chalk Requires pneumatic pick or similar appliance to excavate
Class 2. Compact gravel or sand subsoil	Requires a pick for excavation; a 50 mm wood peg hard to drive more than about 75 mm
Class 3. Clay and sandy clay	Stiff, cannot be moulded in the fingers and requires a pick or pneumatic spade for removal
Class 4. Clay or sandy clay	Firm, can be moulded with the hand but can be excavated with a spade
Class 5. Sand, silty sand and clayey sand	Loose, can be excavated with a spade
Class 6. Silt, clay, sandy clay and silty clay	Soft, can be moulded with the hand and easily excavated
Class 7. Silt, clay, sandy clay and silty clay	Very soft and squeezes through fingers when squashed

Subsoils are of many varieties and may generally be classified as rock, compact gravel or sand, firm clay and firm sandy clay, silty sand and loose clayey sand; see Table 7.1.

It is possible to erect a wall on rock with little or no preparation, but on all soils it is necessary to place a continuous layer of in-situ concrete in the trench called the 'building foundation'. (The term 'in situ' means cast in place in its permanent position; unlike 'pre-cast' which is made elsewhere, lifted and transported later to the place where it is required for use.) This cast, in-situ concrete is made from Portland cement with coarse aggregate, plus sharp sand or ballast, graded from 40 mm to fine sand, and mixed in the proportion of 1:6.

The design of a foundation is an important subject for the architect's consideration, particularly where the building structure is heavy, with possible concentrated loading, and the ground on which the building rests is of poor load-bearing capacity or is affected by other conditions such as seasonal change. Even buildings of the small domestic and industrial type may require careful site exploration so that a suitable foundation is constructed and the future stability of the building assured. Where buildings are extremely heavy it may be necessary to construct foundations of a special type, such as 'piles', where the site consists of deep beds of soft soil overlaying a hard soil, but it is the intention here to deal only with buildings of the smaller domestic and industrial range.

Foundation types

This chapter will deal with the four main types of foundations:

1. Strip foundations – narrow strip/wide strip foundations
2. Pad foundations
3. Raft foundations
4. Piled foundations.

Foundation design

When a load is placed on soil it is necessary to 'spread' or 'extend' the foundation base in order to ensure stability. This extended or spread foundation is referred to as a 'strip foundation' in the case of a continuous wall structure, or a 'pad foundation' in the case of an isolated pier.

The important loads to be resisted by a foundation are the 'dead loads', or the weight of the building structure, and the superimposed loads, or the weight that may be placed on the building structure. The area or spread of the foundation base should be sufficient to resist the downward thrust or bearing pressure of these combined loads.

The spread of a pad foundation (Fig. 7.2) required for a pillar or isolated pier can be determined by dividing the combined loads by the safe bearing capacity of the soil, viz:

$$\frac{\text{Combined load of pier}}{\text{Safe bearing capacity of soil per square metre}} \quad \text{or} \quad \frac{\text{Total load of pier}}{\text{Safe load on soil}}$$

These calculations have been prepared and published in the current Building Regulations for the more common strip foundations on subsoils

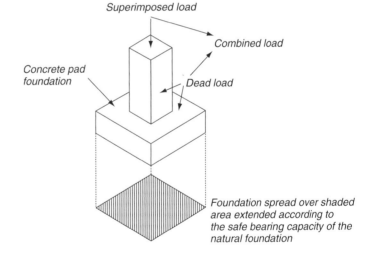

Figure 7.2 Transfer of building load on to subsoil under pad foundation

Superimposed load

Combined load

Concrete pad foundation

Dead load

Foundation spread over shaded area extended according to the safe bearing capacity of the natural foundation

up to a maximum load of 70 kN; see Table 7.2. Any calculations outside this chart should be carried out professionally.

The foundation for a continuous wall structure (Fig. 7.3) is obtained by similar methods; in this case a one metre length of wall is taken for the purpose of calculation.

Having determined the spread of the foundation base it is now necessary to consider the depth. The base, in spreading the load, is subjected to stresses known as 'tension' and 'punching shear' and must therefore be of sufficient depth to resist them. Tension is due to the bending tendency and

Table 7.2 Minimum width of strip foundations

Type of subsoil	Condition of subsoil	Total load of load-bearing walling not more than (kn/linear metre)					
		20	30	40	50	60	70
1. Rock	Sandstone, limestone or firm chalk	In each case equal to the width of the wall					
2. Gravel Sand	Compact Compact	250	300	400	500	600	650
3. Clay Sandy clay	Stiff Stiff	250	300	400	500	600	650
4. Clay Sandy clay	Firm Firm	300	350	450	600	750	850

5. Sand
   Silty clay
   Clayey sand
	Loose
	Loose
	Loose
		400	600				

6. Silt
   Clay
   Sandy clay
   Silty clay
	Soft
	Soft
	Soft
	Soft
		450	650	*Note*: If the total load exceeds 30 N/m then this table does not apply to types 5, 6 and 7			

7. Silt
   Clay
   Sandy clay
   Silty clay
	Very soft
	Very soft
	Very soft
	Very soft
		600	850				

Based on information found in Approved Document A in the current Building Regulations.

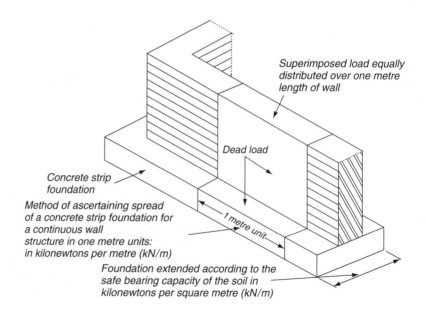

Concrete strip foundation

Superimposed load equally distributed over one metre length of wall

Dead load

Method of ascertaining spread of a concrete strip foundation for a continuous wall structure in one metre units: in kilonewtons per metre (kN/m)

1 metre unit

Foundation extended according to the safe bearing capacity of the soil in kilonewtons per square metre (kN/m)

Figure 7.3 Loading under strip foundation

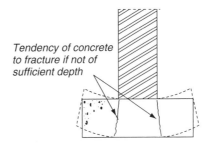

Figure 7.4 Potential failure of a strip foundation due to punching shear

Tendency of concrete to fracture if not of sufficient depth

punching shear to the possibility of the wall or pier structure's tendency to punch a hole through the foundation base (Fig. 7.4).

The thickness of the concrete foundation should not be less than the projection of the strip either side of the wall but in no case less than 150 mm. One method of ascertaining the depth of concrete is shown in Fig. 7.5.

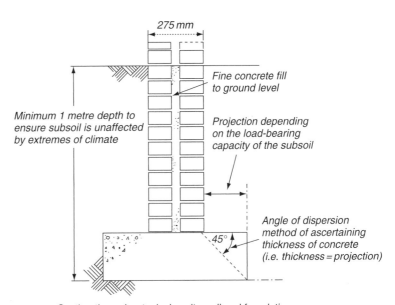

275 mm

Fine concrete fill to ground level

Minimum 1 metre depth to ensure subsoil is unaffected by extremes of climate

Projection depending on the load-bearing capacity of the subsoil

45°

Angle of dispersion method of ascertaining thickness of concrete (i.e. thickness = projection)

Figure 7.5 Establishing thickness of strip foundation concrete

Section through a typical cavity wall and foundation

Design of simple strip foundations

Using the information given it is now possible to calculate the spread and depth of simple concrete foundations.

You require to know the following:

1. The total load per metre run of wall
2. The type of subsoil
3. The width of the wall.

Example 1

Design a strip foundation to carry a 275 mm wide cavity wall, if the total load is 60 kN per metre run of wall and the ground is stiff sandy clay; see Fig. 7.6.

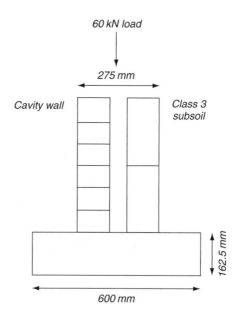

Figure 7.6 Designing a strip foundation

Remember the wall should sit in the middle of the strip foundation.

The first thing to check out is Table 7.2. Find the column with the correct loading of 60 kN and trace down the column until you reach the correct class of subsoil – class 3. The two join at 600 mm. This gives the minimum width of the strip foundation.

The next calculation is to find the projection of the concrete. The wall is 275 mm wide and the total width of the foundation is 600 mm.

If we deduct 275 from 600 and divide it will give the projection:

$$\frac{600 - 275}{2} = 162.5 \text{ mm projection}$$

According to the rules for the projections the depth must be at least equal to the projection but in no case less than 150 mm. Therefore the depth of the foundation concrete should be 162.5 mm.

This is the mathematical method of finding the depth; the angle of dispersion method is shown in Fig. 7.4.

Example 2

Design a strip foundation to carry a 275 mm wide cavity wall, if the total load is 30 kN per metre run of wall and the ground is sandy gravel; see Fig. 7.7.

Again the first thing to check out is Table 7.2. Find the column with the correct loading of 30 kN and trace down the column until you reach the correct class of subsoil – class 2. The two join at 300 mm. This gives the minimum width of the strip foundation.

The next calculation is to find the projection of the concrete. The wall is 275 mm wide and the total width of the foundation is 300 mm.

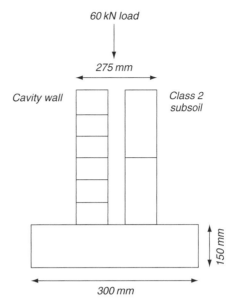

60 kN load

275 mm

Cavity wall

Class 2
subsoil

150 mm

300 mm

Figure 7.7 Designing a strip foundation

If we deduct 275 from 300 and divide, it will give the projection:

$$\frac{300 - 275}{2} = 12.5 \text{ mm projection}$$

According to the rules for the projections the depth must be at least equal to the projection but in no case less than 150 mm. Therefore the depth of the foundation concrete in this example should be 150 mm.

These narrow foundations, although acceptable by the local authority, are practically impossible to build. They are usually extended in width up to 150 mm to allow the bricklayer to stand in the trench when foundation walling is being built.

Atmospheric depth

This is the depth below ground level to which foundations should be taken.

It depends on the type of soil and is the depth at which the sub-foundation ceases to be affected by the weather. This is between 600 mm and 1500 mm, decreasing as the proportion of gravel increases (Fig. 7.5).

In buildings of the smaller domestic and industrial type, or where the loads are not excessive, standard methods and rules are applied to the design of the foundations. Generally, the resultant construction complies with the Building Regulations, but in all cases the decision as to adequacy rests with the local authority building control office.

Other foundation types

Narrow strip

As in the last example, the working space required to build on top of the concrete strip foundation would make the strip wider than it would need to be to carry the load. In these circumstances, an economical alternative is the 'narrow strip' or 'trench fill' as it is sometimes known; see Fig. 7.8.

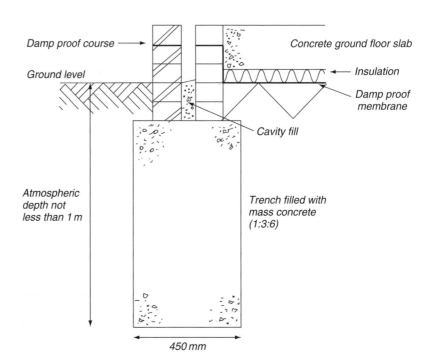

Damp proof course →

Concrete ground floor slab

Ground level

Insulation

Damp proof membrane

Cavity fill

Atmospheric depth not less than 1 m

Trench filled with mass concrete (1:3:6)

450 mm

Figure 7.8 Narrow strip foundation

A narrow strip is excavated by the mechanical excavator and back-filled with mass concrete up to a level just below the finished ground level.

A high standard of accuracy in constructing such a foundation is required.

It is cheaper and quicker to fill the trenches with mass concrete than to excavate a wider trench. There is less excavated material to be removed and backfilling is eliminated.

This trench for a strip foundation would require timbering for safety purposes, foundation concrete would require laying, the foundation brickwork constructed and the cavity filled with weak concrete up to ground level. This method could be reduced by using foundation blocks, but would still work out more expensive than trench fill.

Wide strip foundations

Where the structural loads are very heavy or where the safe bearing capacity of the soil is low, the spread of the foundation base would become greater and this is normally referred to as a 'wide strip foundation'. It follows that the required depth of the concrete foundation base may be considered excessive and it can be reduced by the introduction of steel reinforcement, but the foundation must always be of sufficient depth to ensure that, in combination with the steel, it will resist the stresses of tension and shear. Figure 7.9 shows a simple example of a reinforced concrete strip foundation.

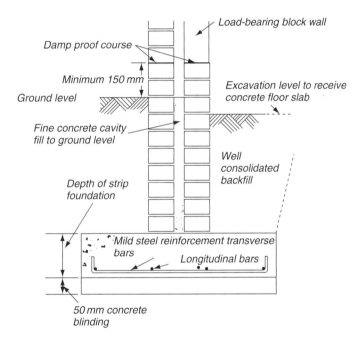

Figure 7.9 Reinforced wide strip foundation

Pad foundations

Loads are not always evenly distributed along the wall, but may at times be concentrated at various points. Figure 7.10 shows part of a wall with an attached pier, which may, for instance, bear a roof truss or beam. In order to obtain even pressure over the soil the foundation would be extended as shown.

For single loads which are transmitted down a column, the most common foundation is a square or rectangular block of concrete of uniform thickness known as a 'pad' foundation; see Fig. 7.11.

It is sometimes more economical to construct a foundation of isolated pads with pillars of brick or concrete, which in turn would support concrete ground beams and concrete floor slab, which in turn would then support the walls of the building.

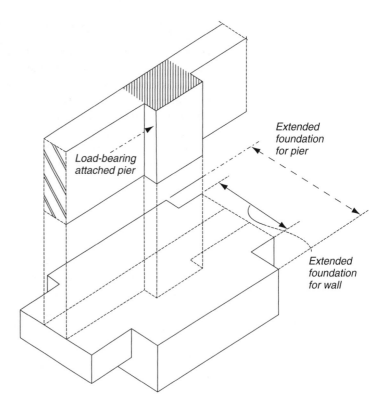

Figure 7.10 Extension of strip concrete foundation around attached pier

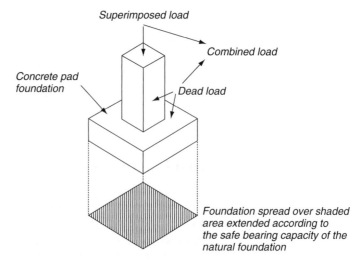

Figure 7.11 Pad foundation

This method of construction would save total trench excavation, timbering to the trenches and foundation brickwork all the way around the perimeter of the building.

In order to spread the load over a greater area it is necessary either to make the pad thicker or use reinforced concrete.

Raft foundations

These foundations consist of a raft of reinforced concrete under the whole of the building. They are often used on poor subsoils for lightly loaded buildings and are considered capable of accommodating small settlements of the subsoil.

The simplest and cheapest form of raft is the thick reinforced concrete raft; see Fig. 7.12. Its rigidity would enable it to minimise the effects of differential settlement.

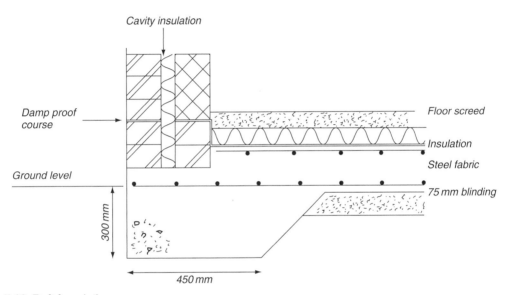

Figure 7.12 Raft foundation

Short bored piled foundations

If instead of spreading the load from the wall over a wide area, it is decided to transfer it to a greater depth, an economical solution is the use of a 'short bored pile' foundation.

Short bored piles are formed by boring circular holes of 300 mm diameter to a depth of about 3 m by means of an auger. This depth is governed by the level of suitable bearing capacity ground. These are filled as soon as possible with mass concrete.

The piles are placed at the corners of the building and at intermediate positions along the walls.

The piles support reinforced concrete beams which are cast in place in the ground; see Fig. 7.13.

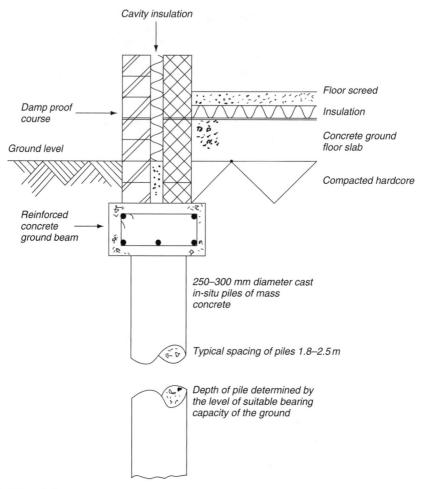

Figure 7.13 Piled foundation

Brick footings

Due mainly to the quality of concrete now used in modern construction prac-tice, many authorities consider brick footings to be obsolete. However, it is felt that the apprentice should be aware of the principles involved when this form of construction is adopted and of the necessary bonding arrangements.

It is possible to ease the stresses of tension and shear in a foundation base by the addition of a construction known as a 'footing' and where brick is used this is achieved by regular offsets at the base of the wall or pier structure; the footings spread the weight of the wall or pier and superimposed loads over the concrete, which in turn distributes the combined loads over the soil.

Thus the first course of footings is always double the wall width, the second and each following course of footings is offset 56 mm each side. Every course of footings should be header bond and sectional, following Rule of Bonding 7; see Fig. 7.14.

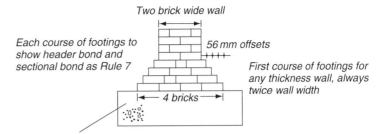

Two brick wide wall

Each course of footings to
show header bond and
sectional bond as Rule 7

56 mm offsets

First course of footings for
any thickness wall, always
twice wall width

4 bricks

Figure 7.14 Cross section of footing courses to a two-brick wall

Strip foundation in Victorian or Edwardian age
buildings may not be concrete but rammed
'hoggin' instead (that is 'as-dug' gravel,
before clay has been washed out)

Stepped foundations

These are constructed on sloping sites in order to ensure a horizontal bearing on the natural foundations (Fig. 7.15). Note the overlap of concrete at the change of levels. The height of the steps must be maintained at not more than 450 mm and where this dimension is exceeded special precautions may be necessary.

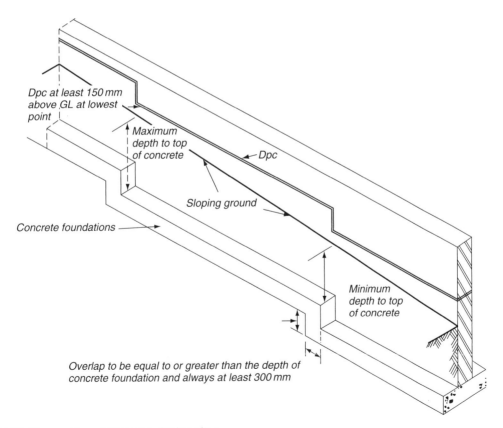

Dpc at least 150 mm
above GL at lowest
point

Maximum
depth to top
of concrete

Dpc

Sloping ground

Concrete foundations

Minimum
depth to top
of concrete

Overlap to be equal to or greater than the depth of
concrete foundation and always at least 300 mm

Figure 7.15 Stepped foundations on sloping sites

8 Craft operations

Bricklaying

The bricklayer has no conventional rules to work with, but the rules to be used are those which are the outcome of experience.

The apprentice will no doubt have other problems to handle besides those shown. Time should be spent to visualise the job, consider the best methods of approach, and use all the skills possible, whether the work is to be covered or is to be highly decorative, for all to see. As a learner, the apprentice should never allow quality to be sacrificed for speed, which will be attained by constant practice.

The stability and appearance of the work should always be the master craftsman's chief concern.

Good bricklaying entails the ability to master the art of spreading the mortar bed, dexterity in handling the brick to be laid, and the possession of a keen eye. All these can be acquired by practice. Before laying bricks on any job, place the mortar or 'spot' board, with the bricks or blocks, in a convenient position. They must be within easy reach so that no unnecessary movement is involved when materials are required. Block up the spot board on bricks, one at each corner, so that it is kept clean, and load out as shown for a one-brick wall, or a cavity wall (Fig. 8.1).

When loading out always take facing bricks from as many packs as possible, but at least three, this will avoid banding of the bricks.

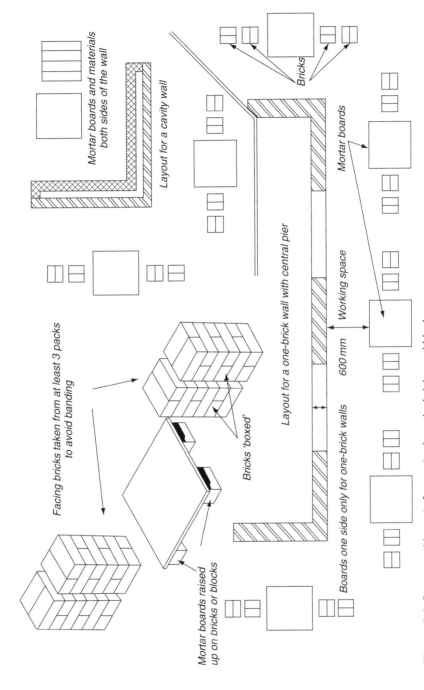

Mortar boards and materials both sides of the wall

Layout for a cavity wall

Bricks

Mortar boards

Facing bricks taken from at least 3 packs to avoid banding

Bricks 'boxed'

Layout for a one-brick wall with central pier

Working space

600 mm

Mortar boards raised up on bricks or blocks

Boards one side only for one-brick walls

Figure 8.1 Suggested layouts for mortar boards, bricks and blocks

Do not grasp the trowel as if clenching the fist, but place the thumb on the ferrule and handle lightly, so that a flexible wrist action is possible (Fig. 8.2). Pick up the mortar with an easy sweeping motion and spread it on the wall sufficiently thick to allow the brick to be placed by pressure of the hand. A common fault is the placing of too much mortar under the brick, so that considerable hammering and tapping are necessary before the brick reaches its final position. The bricklayer usually estimates the amount of mortar bed required by the feel of the brick.

Figure 8.2 Method of holding the brick trowel

Bricks of the hand-made type are often slightly misshapen and some difficulty may be experienced in keeping a flat face and preventing 'hatching and grinning' (Fig. 8.3(a)). If bricks are cambered in their length, lay them as shown in Fig. 8.3(b). Never allow them to 'cock up' at the back (Fig. 8.3(c)), as this makes the laying of the next course difficult and tends to place the wall out of level in its width. The method shown sometimes necessitates the laying of the brick frog downwards and the filling of the frog with mortar before laying, to maintain the solidity of the wall. In ordinary circumstances, however, bricks should always be laid with the frog upwards.

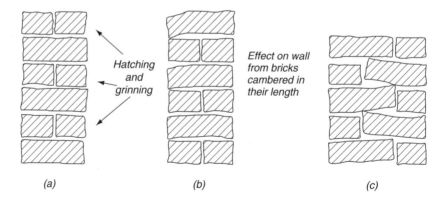

Figure 8.3 Dealing with the natural characteristics of hand-made and stock bricks

(a)   (b)   (c)

Engineering bricks are always difficult to lay, as they have a tendency to 'swim' owing to their density and non-absorptive nature.

When laying these bricks, see that the mortar is as stiff as possible for easy handling, lay the bricks in the required position and leave them, as any attempt to touch them after laying will create difficulties. Have confidence when handling them and do not resort to tricks such as sprinkling the mortar bed with neat cement or placing absorptive bricks on the mortar bed to withdraw the moisture before laying engineering bricks. Both are unnecessary – the former is costly, while the latter tends to destroy the hardening action of the cement mortar.

In hot summer weather bricks, other than those of the engineering type, should be wetted to wash off surplus dust and to prevent undue absorption of moisture from the mortar bed. In winter months this is not necessary as the atmosphere is usually sufficiently damp to achieve these purposes. At this time of the year, however, brickwork should be protected overnight against frost. After the day's work is completed and before leaving the job, the last course of brickwork should be covered with hessian sacking or other suitable material which may be available.

Face work, too, becomes stained through rain splashing back from the scaffold or from the platform from which the bricklayer works. This can be prevented by turning the front scaffold board on edge, as shown in Fig. 8.4.

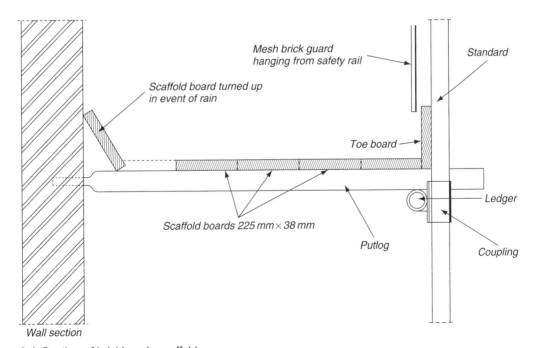

Figure 8.4 Section of bricklayer's scaffold

Erecting a brick wall

When building a wall over 1.125 m in length it is always advisable to use a line and pins, for this will ensure a neat job. Before bringing these into use, it is first necessary to build the corners, making sure that they are vertical or 'plumb', in alignment (Fig. 8.5), and to 'gauge'.

As each quoin brick is laid it should be plumbed with the aid of a spirit level and checked for gauge with a tape or gauge rod.

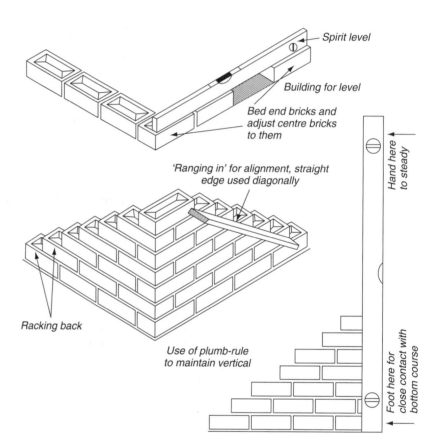

Spirit level

Building for level

Bed end bricks and adjust centre bricks to them

'Ranging in' for alignment, straight edge used diagonally

Racking back

Use of plumb-rule to maintain vertical

Hand here to steady

Foot here for close contact with bottom course

Figure 8.5 Raising quoins in half brick walls

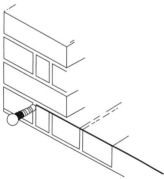

Figure 8.6 Use of line pins

If we assume the building of a straight one-brick wall, first, erect approximately six courses of brickwork on the corners and then begin to use a line and pins. Two ways of keeping the line taut are shown. Figure 8.6 shows a method commonly used – the placing of the line pin in a vertical joint. Figure 8.7, showing the use of corner blocks, illustrates the better method, especially where expensive facing bricks are being used.

When laying bricks to a line, always ensure that a trace of daylight can be seen between the line and brick (Fig. 8.8). This prevents the laying of the bricks 'hard' to the line, which if continued would eventually

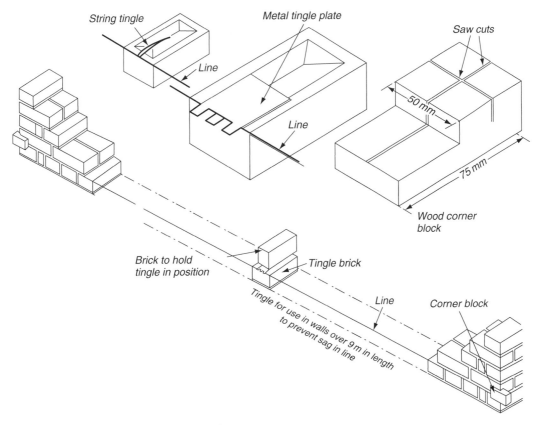

Figure 8.7 Use of tingle plate and corner blocks

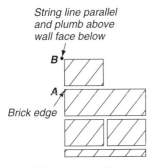

Figure 8.8 Laying to line

place the wall out of alignment to a considerable extent. To keep the wall flat and to prevent hatching and grinning, imagine the bricks are being laid between two lines, one being the edge of the previous course laid to a line 'A' (Fig. 8.8) and the other to the present string line position at 'B'.

If a wall exceeds 9 m in length, it is necessary to use a 'tingle' to take the sag out of the line. The tingle brick should be as near the centre of the wall as possible, and must be sighted through from corner to corner every time the line is raised one course, to ensure that the string line is being supported at the correct level (see the bottom right-hand end of Fig. 8.7). A tingle achieves its purpose up to a wall length of approximately 12 m. Beyond this, it is advisable to divide the wall into two parts and to erect part of the wall in its centre to act as a corner. This should be located at a vertical movement joint or be beneath a pier, and is termed a 'lead' (Fig. 8.9). The wall illustrated is considered to have been erected, in order to show clearly the positions of the 'lead' and tingle.

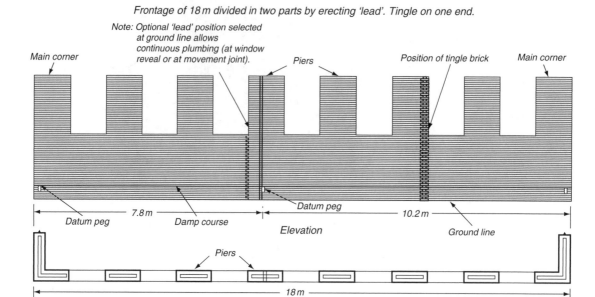

Figure 8.9 Erecting a long wall

Raising quoins

The first course should be laid dry to ascertain the correct bonding and the position of the quoin bricks. Lay the quoin brick first to gauge and level. This brick is then used to gauge and plumb from so it is important that it is correct.

When deciding how high to make the quoin remember that every brick in length means one course in the quoin. Therefore for a small quoin seven courses high you need to lay three bricks along one wall and four bricks along the other; see Fig. 8.5.

Corners should never be built with a straight line of toothings, as in Fig. 8.10, as it is difficult to make good the toothings in all cases. The correct method is to build, as shown in Fig. 8.11, continually racking back and maintaining alignment of the corner by means of a profile. A combination of racking back and toothing is shown in Fig. 8.12. This prevents a direct line of toothings and avoids excessive racking out of brickwork.

Corners must always be built before running in the wall so that there is always somewhere to fix line and pins or corner blocks throughout the day. It is preferable if corners are only raised a few courses at a time, say only six or seven courses ahead of the line, to avoid the quoin courses getting out of face plane alignment with the main walling.

Quoins must be large enough to resist the pulling power of the line each time it is tightened up. They must also be of just sufficient height to allow accurate ranging-in with the spirit level as indicated in Fig. 8.5.

However, if it is necessary for some other reason to build a large quoin, as shown in Fig. 8.11, then a temporary timber profile must be set up as indicated, so as to ensure that the racking-back courses will be truly in line with the overall wall face.

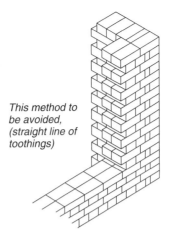

This method to be avoided, (straight line of toothings)

Figure 8.10 Bad practice

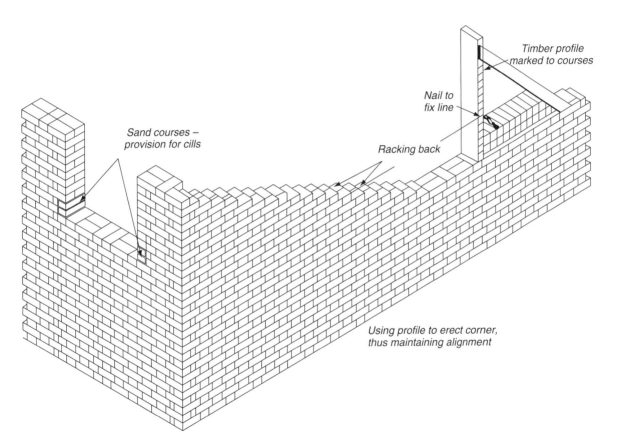

Timber profile marked to courses

Nail to fix line

Sand courses – provision for cills

Racking back

Using profile to erect corner, thus maintaining alignment

Figure 8.11 Raising a large quoin

Figure 8.13 shows the method of erecting a wall, with a number of attached piers. In order to maintain correct levels, it is always advisable to lay the course of brickwork on the wall before attempting to lay any bricks on the piers.

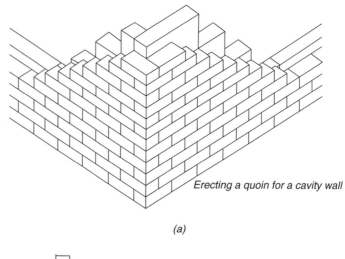

Erecting a quoin for a cavity wall

(a)

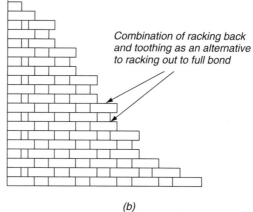

Combination of racking back and toothing as an alternative to racking out to full bond

Figure 8.12 More convenient sized alternative to Fig. 8.11

(b)

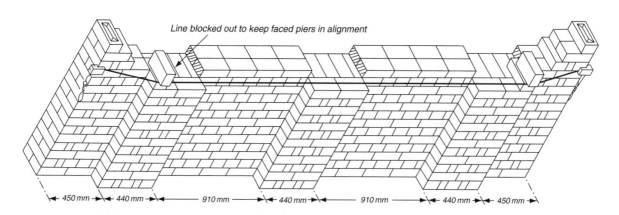

Line blocked out to keep faced piers in alignment

|← 450 mm →|← 440 mm →|← 910 mm →|← 440 mm →|← 910 mm →|← 440 mm →|← 450 mm →|

Note: Actual dimensions of attached piers and intermediate spaces are shown, which will apply when setting out instead of nominal dimensions of 450 mm and 900 mm respectively.

Figure 8.13 Erecting a wall with an attached pier

There are many patent profiles available now which allow the brick-layer to run the line without setting up a quoin. These should be erected first at both corners, checked for plumb and gauge and lines attached. The wall is then run without the need to build the corners first. Versions are available as profiles for openings, cutting up gables etc.

Preliminary setting-out of a building

Before any setting-out should be undertaken the topsoil should be removed. It is assumed that the building of a traditional semi-detached domestic dwelling is contemplated (Fig. 8.14), but it should be noted that the same principles can be applied to all types of buildings and other more compli-cated forms.

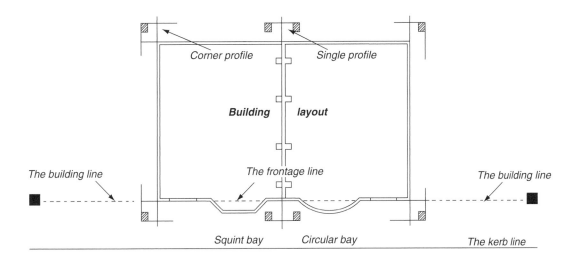

Figure 8.14 Location of foundation profiles

When the setting-out stage is reached the site would have been previously surveyed and probably levelled to find the shape and features of the land.

Drawings showing the proposed buildings and associated site works will have been submitted to the local authority and approved.

A large development requires setting-out drawings to be prepared by the architect or engineer, but a small site would have sufficient informa-tion on the block plan or general site plan.

The most important task in setting out a building is to establish a base line to which all other setting-out can be related to. This base line is very often the building line. The building line is an imaginary line which is established by the local authority. You can build on the building line and any distance behind the building line but:

NEVER BUILD IN FRONT OF THE BUILDING LINE

It is usual for the building line to be given as a distance from either:

* existing buildings;
* the centre line of the road; or
* the kerb line.

The frontage line of the proposed building must then be set out on or behind the building line, but never in front of it. In this example the frontage line is on the building line.

Degree of accuracy

British Standard: 1978 states that the permissible deviation for horizontal brick walls up to 40 m in length is ±40 mm. This required accuracy can be achieved if a steel tape is used and supported to avoid any sag. Checks should be made on all measurements wherever possible.

All horizontal measurements should be made horizontal. Use a level to ensure the tape is horizontal to avoid errors; see Fig. 8.15.

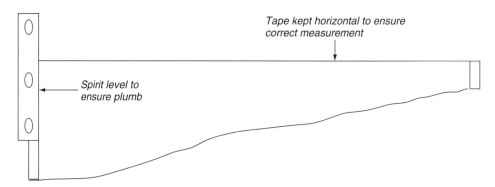

Figure 8.15 Accuracy with horizontal measurements

The building line is fixed to two pegs on either side of the proposed plot, a reasonable distance away from the building. The two corner pegs are then set out on the building line which marks out the frontage line of the proposed building; see Fig. 8.16.

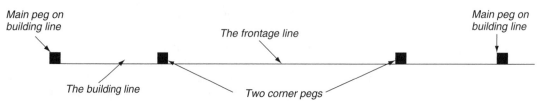

Figure 8.16 Setting out frontage line

The side walls of the building can now be set out at right angles to the frontage line. Ensure the pegs are placed at a distance more than the side wall length; see Fig. 8.17.

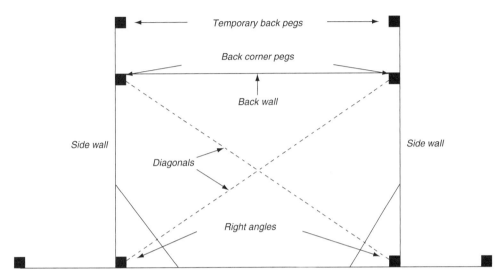

Figure 8.17 Checking for accuracy

The right angles can be set out with a builder's square, the 3:4:5 method or an optical square.

Once the position of the side walls has been established the correct length can be measured and pegs can be knocked in. The back line can now be completed. Once the outline has been completed it is essential that the setting-out is checked for square. The two diagonals are measured and if they are equal the setting-out is accurate and the building is square; see Fig. 8.17. If they are not equal adjustments have to be made. Remember, never alter the frontage line as this should be correct. The back two pegs have to be adjusted to ensure the diagonals are equal. When adjustments have been made always recheck the dimensions of all wall lengths.

When the setting-out has been proven correct the profiles have to be erected. At the moment the setting-out pegs would be in the foundation trench so they have to be repositioned approximately one metre away from all the wall faces. Profiles can be either corner or single type and consist of wooden pegs and rails; see Fig. 8.18.

Profiles are erected to enable the corner setting-out pegs to be removed to allow excavation to take place without disturbing the pegs.

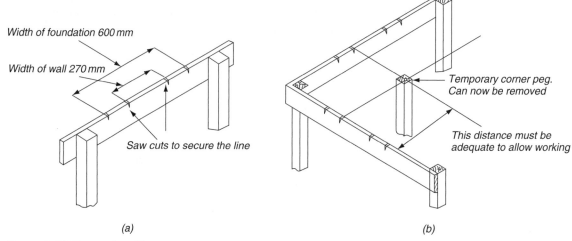

Width of foundation 600 mm

Width of wall 270 mm

Saw cuts to secure the line

Temporary corner peg.
Can now be removed

This distance must be
adequate to allow working

(a)   (b)

Figure 8.18 Types of profiles

One of these corner profiles could be set to a given level which is
known as the 'datum' and it may relate to the finished floor level or damp
proof course. This datum peg should be protected with concrete.

The datum peg could also be positioned away from the setting pegs but
close enough to be accessible and should also be protected with concrete
and a small barrier of pegs and rails; see Fig. 8.19.

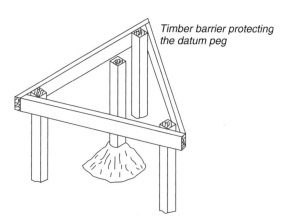

Timber barrier protecting
the datum peg

Figure 8.19 Datum peg

The profiles should be positioned approximately 1 metre away from
the face of the building to allow working space for the excavation.

The completed profiles can have foundation and wall widths marked
on them in one of various methods. Saw cuts are best as nails could be
accidentally removed; see Fig. 8.20.

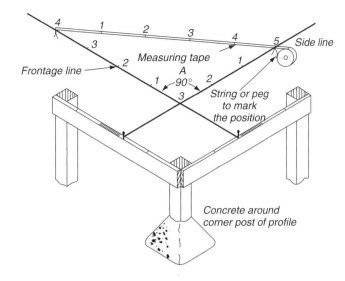

Figure 8.20 Checking a right angle by the 3:4:5 method

Once the profiles have been accurately constructed the dimensions should all be checked again, as should the diagonals.

The original setting-out pegs can now be removed.

Building lines can now be fastened to the profiles and the trench can be marked out ready for excavation as shown in Fig. 8.21.

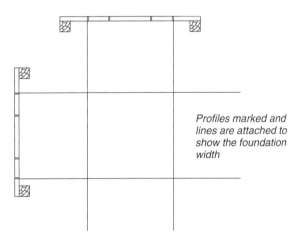

Figure 8.21 Marking out the ground ready for excavation

Timber templates

Framed timber templates can be prepared for the setting-out bays. They are applied as shown in Fig. 8.22.

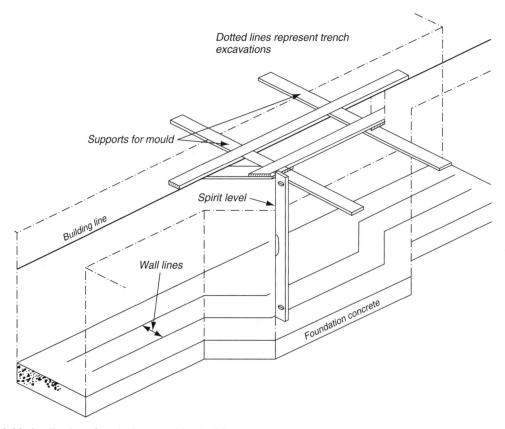

Figure 8.22 Application of squint bay mould to building line

Two methods can be adopted for the construction of circular bays:

1. The use of a framed timber mould for original setting-out, and circular templates for building. This is the usual method, especially where a number of bays are to be built (Fig. 8.23).
2. The use of a trammel (Fig. 8.24).

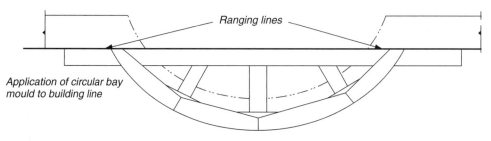

Figure 8.23 Use of segmental bay mould

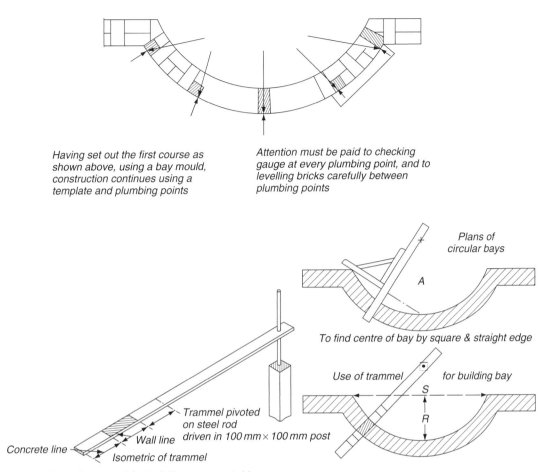

Having set out the first course as shown above, using a bay mould, construction continues using a template and plumbing points

Attention must be paid to checking gauge at every plumbing point, and to levelling bricks carefully between plumbing points

Plans of circular bays

A

To find centre of bay by square & straight edge

Use of trammel *for building bay*

S

R

Trammel pivoted on steel rod driven in 100 mm × 100 mm post

Wall line

Concrete line

Isometric of trammel

Figure 8.24 Use of trammel for building segmental bay

Work below ground level

Almost all construction sites will involve some form of excavation. Every year many construction workers have accidents involved with excavations. The most common problem is the collapse of trench sides or equipment or materials falling into trenches. Any groundwork must be planned and carried out correctly and safely to prevent accidents.

All subsoils vary in their abilities to remain stable during excavations.

The first operation is the removal of the topsoil which should have been completed before the building was set out. This also makes the marking out of the ground easier.

Topsoils can vary in depth from approximately 150 mm to 300 mm, and they contain various organic material which is unsuitable to build on. They could be either removed from the site or stored for later use when landscaping.

The foundation trenches can now be excavated by either hand or machine. Hand excavation is only carried out where the site is restricted to machinery such as small extensions at the rear of existing properties.

Timbering

All subsoils have a natural angle of repose which allows the subsoil to rest unless given support. As the trench is excavated temporary support is placed in the trench according to the type of subsoil. Some simple forms for shallow trench excavations are shown in Fig. 8.25(a), (b) and (c).

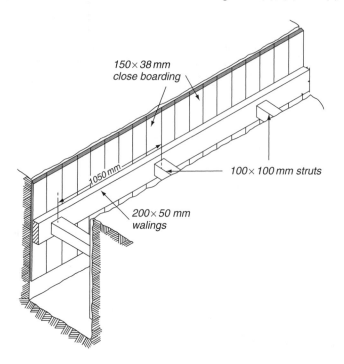

Figure 8.25(a) Loose soil

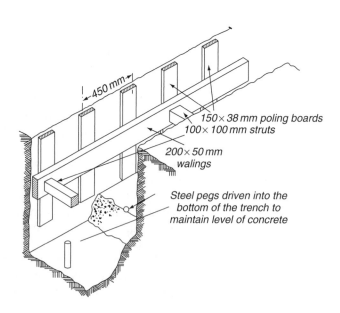

Figure 8.25(b) Moderately firm soil

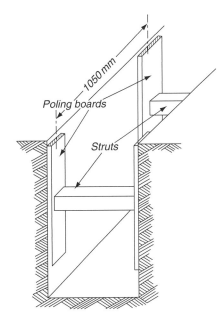

Figure 8.25(c) Firm soil

Laying concrete foundation

After the trenches have been excavated to the required depth the concrete can be placed to form the foundation. A straight edge should be used between each steel peg to produce a level foundation on which to build; see Fig. 8.26.

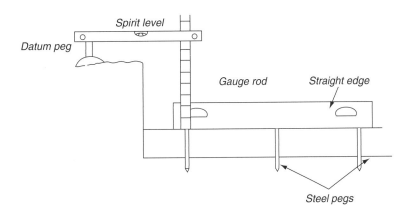

Figure 8.26 Placing concrete in foundations

A rough finish is required to ensure the brickwork adheres to the concrete foundation. The distance from the top of the datum peg to the concrete needs to be in brick courses. This is checked as shown in Fig. 8.26 with a spirit level and gauge rod.

The pegs are then levelled around the trench with either a straight edge or a spirit level. On larger areas, or where more accuracy is required, there are numerous mechanical levels available such as the quickset level, auto set level and even the laser level.

Brickwork up to dpc

Once the concrete has hardened the brickwork can be built. Lines have to be fastened to the profiles to mark out the position of the brickwork; see Fig. 8.27.

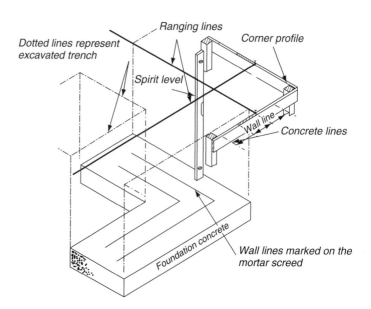

Figure 8.27 Plumbing down from ranging lines

The ranging lines are set to wall marks on the profiles, and plumb lines are taken down on the main corners to the concrete foundation. To facilitate clear marking, a mortar 'screed' must be spread over the concrete to a thickness of approximately 3 mm at the positions where the plumb lines are being taken. A spirit level can normally be used to ascertain plumb lines, but if the trench is too deep to allow this to be done, a 'drop bob' should be employed (Fig. 8.27).

The bricks are laid as explained previously for normal brickwork and the corners are erected by racking back until the correct height is reached.

The first course should be dry bonded to ensure the correct bonding pattern is maintained. The top two courses below the damp proof course should be facing bricks as they will be seen after the ground work has been completed.

Once the corners have been erected the main walls can be completed using lines and pins.

The procedure for working below ground level is very different to working above. The space in the trench may be very tight and all materials have to be set out on the sides of the trenches and the work is mainly below – known as nose bleeding – as you are mainly bending down.

It is important not to place the materials too near the side of the trenches as to cause collapse due to too much pressure (Fig. 8.28).

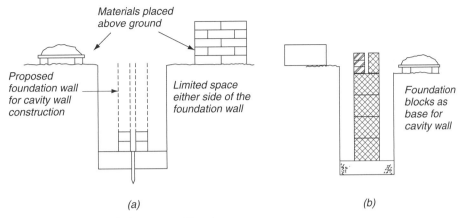

Figure 8.28 Preparation for work below ground level

If the foundation brickwork is a cavity wall there will be more space for the feet while erecting the face wall. Solid brick or foundation block walls will be tight for space as most foundations are designed with a minimum 150 mm space either side of the walls.

From this point, site concrete and damp proof courses are placed. These are described in Chapter 11, but at this stage it is sufficient to know that the ground-floor level is approximately 225 mm above damp proof course level. When this point is reached in the building of the wall, a level must be transferred from the datum which represents ground-floor level. This level will have been adhered to while constructing the wall, to ensure that the wall is horizontal at all points. The transferred level is maintained on the wall by fixing a short length of 50 mm × 25 mm batten, termed a 'datum peg' (Fig. 8.29). From this datum a 'storey rod' is used to maintain correct heights and, in order to be of assistance to the chargehand bricklayer, it

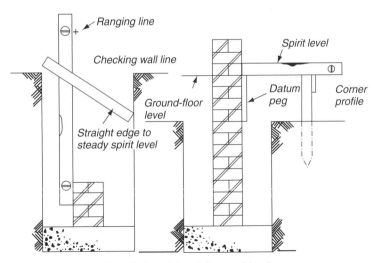

Figure 8.29 Transferring setting-out lines and datum to substructure

Foundation brickwork for a solid wall

should be marked with all those details appearing on the elevation of the wall, e.g. sills, arch and floor levels, etc. (Fig. 8.30).

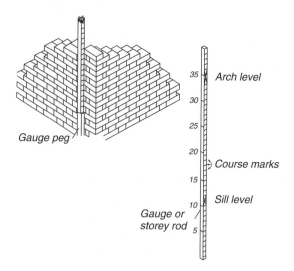

Figure 8.30 Use of storey rod

Services

During work below ground level there are times when openings have to be left in the brickwork for services etc.; see Fig. 8.31. Sometimes the exact position is not known so sand courses are used to allow bricks or blocks to be removed at a later date.

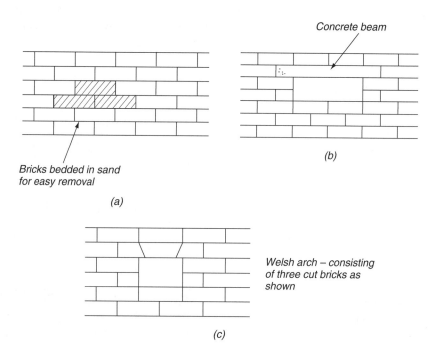

Figure 8.31 Provision for services

Setting out of window openings

Immediately a wall appears above ground level, preparation must be made in the bond arrangement for the insertion of window openings at a future height. At this stage, the proposed elevation of the building must be thoroughly studied and the work visualised so that every feature is known subconsciously, together with its correct position.

In common brickwork which is to be covered, no special care is needed in the bond arrangement, provided all perpends are plumb, but where bricks form the face of a building, special attention must be paid to the bond arrangement.

Whatever the type of bonding used, it must be arranged so that no straight joints occur when window openings are set out and the brick piers formed.

Figures 8.32 to 8.34 show a traditional design form; the dimensions of the windows and piers have been chosen to illustrate the procedure that should be followed in order to present a good appearance of the face brickwork.

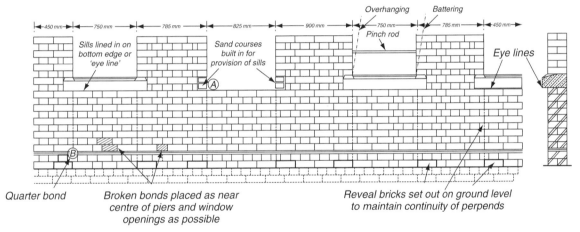

Figure 8.32 Setting out window openings (English bond)

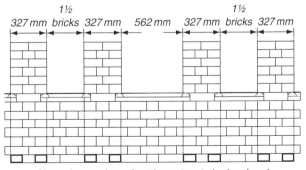

Figure 8.33 Setting out window openings (Flemish bond)

Normal procedure of setting out; note broken bond

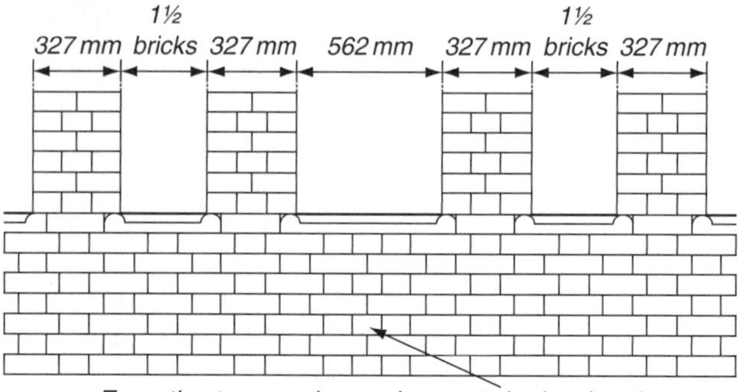

Figure 8.34 Exception to Fig. 8.33 preserving the maximum face bonding. (*Note*: only one broken bond)

Exception to normal procedure; note broken bond

Figure 8.32 shows a complete elevation with the bond set out correctly. In this case a stretcher appeared immediately above ground level; if a header course had occurred it would have been arranged as at 'B' (Fig. 8.32). Note the quarter bond has been allowed for. The probable bonding below ground level is shown by dotted lines. It may be held that the correct bonding could have been arranged at footings level, but the bricks below ground level are often of a different type, and, in addition, the confined area of a trench makes work difficult, and precise setting-out is not always possible.

The positioning of a broken bond will be obvious (see Chapter 4).

The bedding of artificial stone sills, which are built in as the building of the walls proceeds, is also shown. The sills must be bedded at their ends only, as if they are bedded throughout their length, the slightest settlement of the piers would cause them to fracture at the centre. The open joint can be made good when building is completed. When placing sills for alignment, always line or sight along the bottom edge; this is the 'eye line'. Bed the end sills and line in all others to these. Where sills are not built in as the work proceeds, provision is made for their future insertion by building in 'sand courses', as shown at 'A' (Fig. 8.32).

'Pinch rods' (Fig. 8.32) are always useful as they allow a check to be kept at all times on the correct sizes of window openings. These openings should never take the shape shown by the dotted lines, i.e. one reveal 'battered' and the other 'overhanging'. Pinch rods are not needed if window frames are built in.

Figure 8.34 shows an exception to the usual method of setting out window openings. Examination of the two drawings will reveal that if the normal procedure is adopted (Fig. 8.33), four broken bonds occur, whereas if the problem is fully considered before the work is begun, a precise arrangement of bricks will require the use of one broken bond only.

Block bonding

A method adopted to bond a new wall into an existing building. Three types of work are shown:

1. Block bonding a cross wall (Fig. 8.35)
2. Extending the length of a wall (Fig. 8.36)
3. Thickening an existing wall (Fig. 8.37).

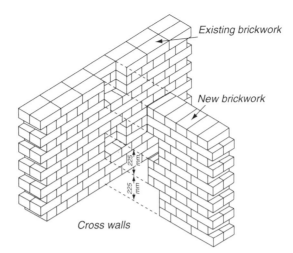

Figure 8.35 Block bonding between walls at right angles

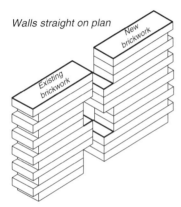

Figure 8.36 Block bonding to lengthen a wall

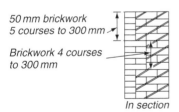

Figure 8.37 Block bonding 50 mm bricks to standard gauge brickwork

Figure 8.38 Block bonding to increase the thickness of the wall

In Figs 8.35 and 8.36 indented toothings could be cut and used for bonding purposes. In some circumstances this is a necessity, but block bonding should be used wherever possible, for two reasons:

1. A clear-cut hole and a more adequate bond are obtained.
2. The existing brickwork is not always of the same gauge as the new brickwork and attempts to bond on alternate courses are impractical.

In Fig. 8.38 block bonding is the most practical solution. Note that the blocks have been placed in a diagonal pattern, which gives greater surface bond than would be the case if the blocks were placed one above the other.

In the illustrations, block bondings have been cut away at every 225 mm. This is the usual practice, but the bondings can be extended to 300 mm without loss of stability.

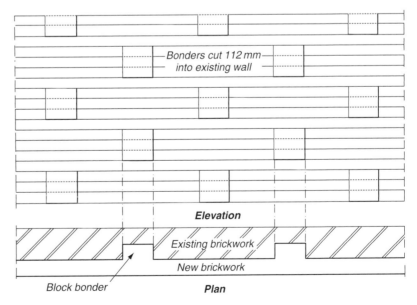

In preparation for work of this type, the bricklayer cuts away the necessary brickwork with a cold chisel and club hammer. There are bricklayers who consider cutting away to be unskilled labour, but this is not the case, for the work does in fact require considerable skill.

Quite often an apprentice employed on this work can be seen attempting to force the chisel, which has been entered into the face of the wall, by using the club hammer with both hands. No doubt expecting to see large pieces of brickwork fall away as a result, and disappointed when the brickwork splays and fractures in the wrong position or when the chisel disappears to its head and is difficult to extract. The art of cutting away requires that:

1. The chisel should never be forced.
2. A start should be made with a small hole, which should be cleared as cutting proceeds and which should be kept symmetrical as it is gradually enlarged.
3. The cutting edge of the chisel, not its head, should be watched.

Figure 8.37 shows the block bonding of a wall in its thickness, when faced with 50 mm facing bricks. Common bricks are not manufactured in 50 mm sizes, and it would be too expensive to use 50 mm facings throughout the thickness of the wall. A combination of both types of brick is therefore needed. A system of block bonding in the section of the wall is adopted to effect this, the blocks occurring at every 300 mm, the wall being level at multiples of this height.

Toothings and indented toothings

It is occasionally found necessary to leave part of a building down and, in preparation for its future extension, toothings or indented toothings are formed as the work proceeds. In order to strengthen the connection between the walls when the extension is built proprietary mesh reinforcements supplied in wall widths are built in (Fig. 8.39).

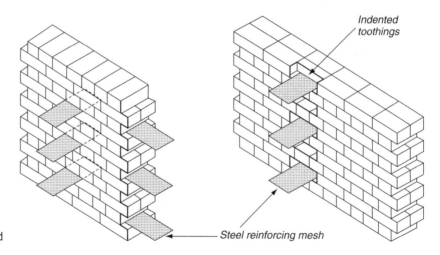

Indented toothings

Steel reinforcing mesh

Figure 8.39 Proprietary mesh used to reinforce toothings or indents and create a mechanical tie for blockwork where indents are not specified

Blockwork

Quoins

Corners should be built as for brickwork. It is important to maintain half bond.

The method most commonly used is to introduce a closer next to the first block, this will allow half bond to be maintained; see Fig. 8.40.

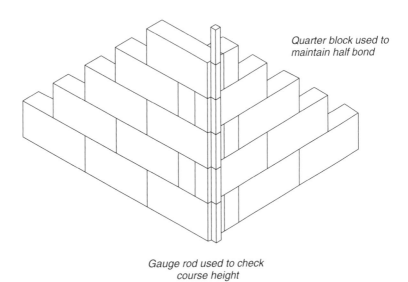

Quarter block used to maintain half bond

Gauge rod used to check course height

Figure 8.40 Quoin built in blockwork

It is important that blocks of the same material and strength are used. Never use bricks to achieve bond!

Check course heights regularly with a vertical gauge rod.

In summer the blocks may be too dry to ensure adhesion with the mortar. It is possible to alter the consistency of the mortar and the mortar should be laid in shorter lengths. **Never dampen down the blocks**.

As mentioned before it is recommended that block walls forming partitions should only be built up to six courses in the day. This will allow the wall to set before building any higher. It is also recommended that supports/profiles could be pre-erected to assist in the building of the wall but also to prevent buckling.

When block walls are erected as part of a cavity wall they are supported with wall ties, therefore profiles are not necessary (Fig. 8.41).

Lintel bearings

The lintel manufacturer's recommendation for minimum bearing should be followed. Lintels should bear on to a full length block shown in Fig. 8.42.

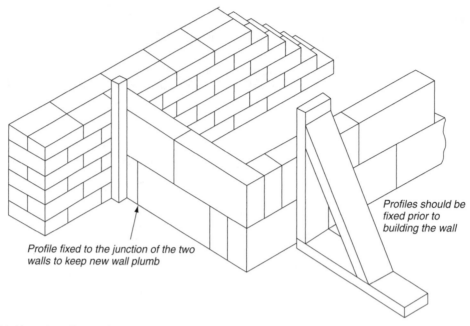

Figure 8.41 Use of profiles on internal blockwork walls

Profile fixed to the junction of the two walls to keep new wall plumb

Profiles should be fixed prior to building the wall

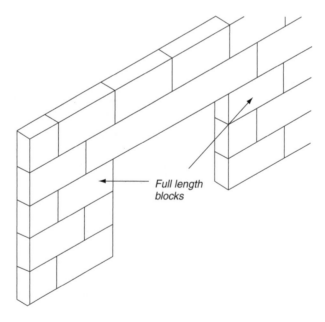

Full length blocks

Figure 8.42 Bearings for lintels

Block bonding

Partition walls can be block bonded to brick walls to gain the maximum stability.

Three courses of bricks are equal to one course of blocks so this is the pattern used to form the block bonding (Fig. 8.43).

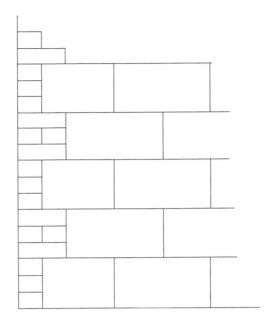

Figure 8.43 Block bonding

Remember always use the same density blocks in the same wall. Never mix bricks and blocks. Most manufacturers supply cut blocks to avoid excessive waste on construction sites. Figure 8.44 shows the various cuts available.

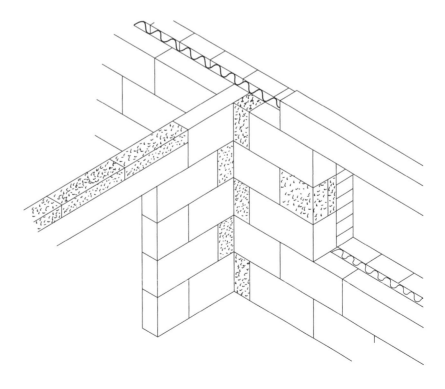

Figure 8.44 Use of cut blocks

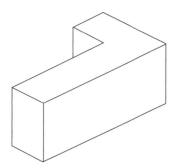

Cavity walls

Openings may be bridged with lintels. The texture of the lintel should match the blocks. When using blocks for cavity walls difficulty may be met when trying to cut small pieces of blocks to close the cavity. Some manufacturers provide return blocks for closing cavities (Fig. 8.45). These are of course more expensive than the ordinary blocks but generally are more economical in time and waste.

Figure 8.45 Special cavity closure

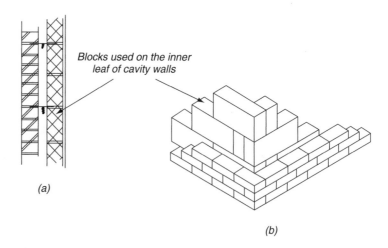

Blocks used on the inner leaf of cavity walls

(a)

(b)

Figure 8.46 Cavity details

Partition walls

One of the most common uses of blocks is in partition walls (Fig. 8.47). These can either be a cavity or solid wall as required by the architect.

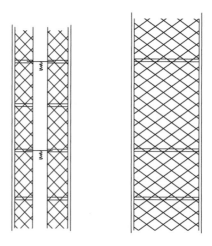

Figure 8.47 Partition walls

Cavity wall construction   *Solid wall construction*

Remember with thinner blocks for partition walls it is essential to provide temporary profiles to support the blockwork until it has set.

Blocks below ground level

Certain quality blocks can be used below ground level; see Fig. 8.48. Always consult the manufacturer's instructions before using.

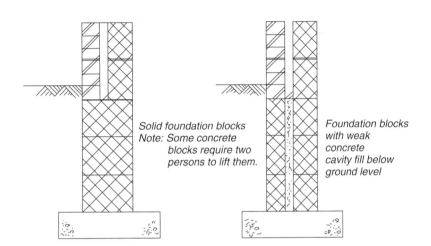

*Solid foundation blocks
Note: Some concrete
blocks require two
persons to lift them.*

*Foundation blocks
with weak
concrete
cavity fill below
ground level*

Figure 8.48 Blockwork below ground level

Decorative panels

Flush panels

Figure 8.49 shows the necessary operations.

1. *Preparation.* The main brickwork is erected and provision is made for the insertion of the panel. The backing is built as the main work proceeds, as this gives adequate bond and assists when the panel is erected.
2. *Erecting the surround.* Tile corners are shown. The plumbing of the side must be watched and line and pins used for the base. Surrounds are not a necessity in this type of panel and have been added in this case as a decorative feature.
3. *Insertion of panel.* Whatever the type of bond adopted, it is always wise to make free use of line and pins, although a straight edge may be used for final adjustment.
4. *Finish.* Make sure that the top of the surround is level with the main brickwork. If it is low it gives an unsightly joint and if it is high unnecessary cutting is involved.

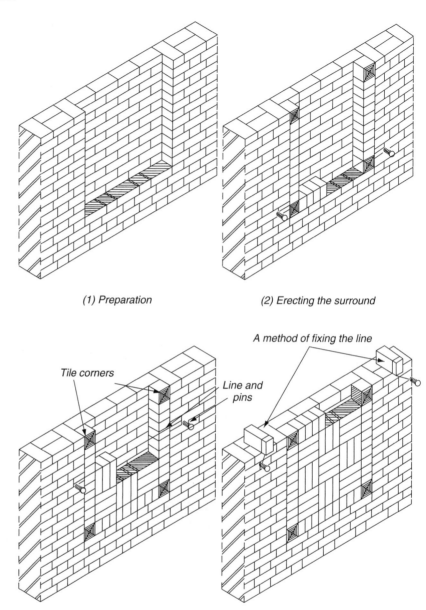

(1) Preparation

(2) Erecting the surround

A method of fixing the line

Tile corners

Line and pins

Figure 8.49 Building a flush panel

(3) Inserting basketweave panel

(4) Finished panel

Projecting panel (Fig. 8.50)

Operations 1 and 3 are similar to those for flush panels. In erecting the surround, partly mitred brick corners have been adopted (Fig. 8.51).

Note that the line and pins have been used on the bottom edge of the surround. This is an eye line. With regard to the finish, a 25 mm projection of the panel leaves a wide joint between backing and panel. A filling of fine concrete is the best practical method of dealing with this, as it

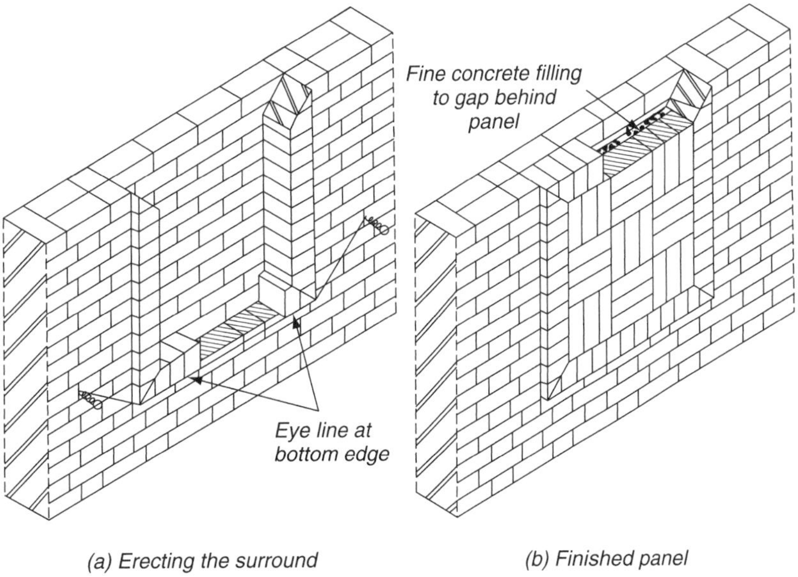

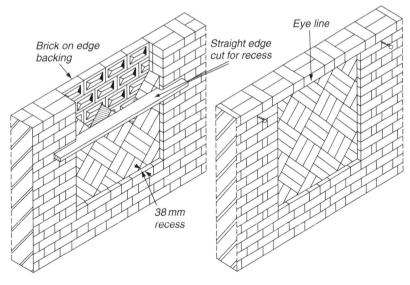

Figure 8.50 Building a projecting panel

(a) Erecting the surround   *(b) Finished panel*

saves excessive use of mortar and obviates the hollow gaps caused by attempting to insert too large a piece of brick.

Recessed panel (Fig. 8.52)

Notice the backing in this case. A one-brick wall is illustrated and the backing is brick on edge, which can be block bonded in every 225 mm. If the wall is thicker than one brick, a brick on edge backing is unnecessary.

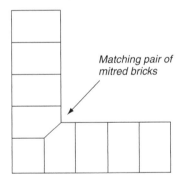

Figure 8.51 Mitred corner to panel surround

Figure 8.52 Building a recessed panel

(a) Inserting the diagonal basketweave panel   *(b) Finished panel*

Inserting the panel

In diagonal basket weave or herring-bone bonds the position of the bottom cuts is important. If the angle is lost it will be found necessary to shorten the length of a brick, which in turn will lead to a search for bricks of excessive length and eventually it may be found impossible to proceed further with the work, owing to the original loss of the angle. If a brick fails to fit into its correct position without cutting when the bottom cuts have been made, the work has been carried out incorrectly and the fault must be found before further work proceeds. When finishing, the eye line must be noted.

Cutting of gable ends (Fig. 8.53)

The illustration is self-explanatory. The correct position of the gallows must be ascertained, both for angle of line and for plumb above the main brickwork. If the latter is correct, the remaining brickwork need not be plumbed. Note the wind-filling between the rafters; this work is usually carried out by the apprentice and is excellent practice in the use of the trowel. It may become monotonous and the apprentice may be tempted to lose interest, but it affords good training and the study of roof construction is an absorbing one.

Cutting brickwork to a lean-to (Fig. 8.54)

The brickwork must be racked back to obtain the correct cut and a fixing for the line and pins. Note the brick corbels supporting the 'wall plates'. These are usually 100 mm × 50 mm timbers, and they are used to distribute the weight of the roof and to obtain a firm fixing. The bricklayer has to bed these wall plates and this must be done solidly and level, using a 3 m to 4 m straight edge of timber or hollow section aluminium and spirit level.

Procedure for raking cutting

1. Rack back and tooth out below the sloping line and pins as shown in Fig. 8.55.
2. Set the sliding bevel at the necessary angle of the raking cutting.
3. Measure the top edge of the cut brick required.
4. Deduct 10 mm from one cross joint.
5. Mark in pencil the remaining measurement on the brick to be cut, back and front.
6. Draw in pencil the angle of sloping (raking) cut, using the sliding bevel, also back and front, taking care to match the slope on the face.
7. Cut the brick with hammer and bolster.
8. Trim the cut face with scutch or comb hammer, slightly 'undercutting' if possible to avoid any projections which would prevent the coping or capping from being bedded properly.
9. Carefully bed the cut brick to line and level.

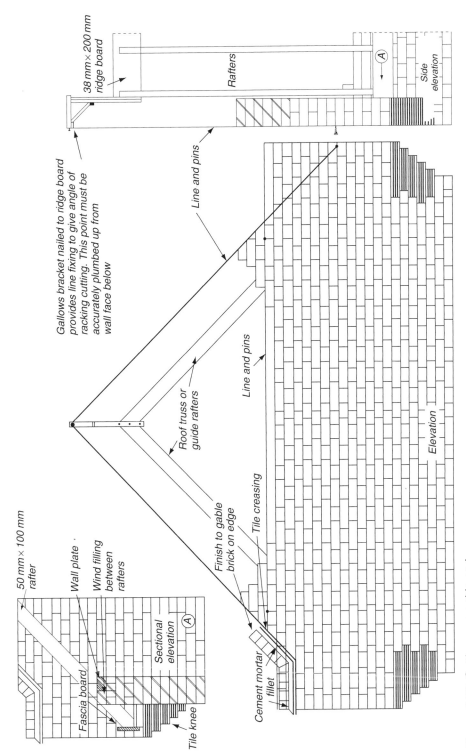

Figure 8.53 Cutting up a gable end

Gallows bracket nailed to ridge board provides line fixing to give angle of racking cutting. This point must be accurately plumbed up from wall face below

38 mm × 200 mm ridge board

Rafters

Side elevation

Line and pins

Roof truss or guide rafters

Line and pins

Elevation

A

Tile creasing

Finish to gable brick on edge

Cement mortar fillet

50 mm × 100 mm rafter

Wall plate

Wind filling between rafters

Fascia board

Sectional elevation

A

Tile knee

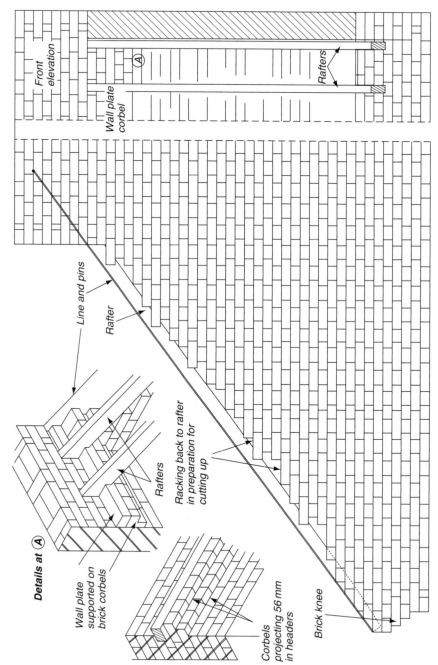

Front elevation

Wall plate corbel

Rafters

Line and pins

Rafter

Racking back to rafter in preparation for cutting up

Brick knee

Details at Ⓐ

Wall plate supported on brick corbels

Rafters

Corbels projecting 56 mm in headers

Figure 8.54 Cutting brickwork to a lean-to roof

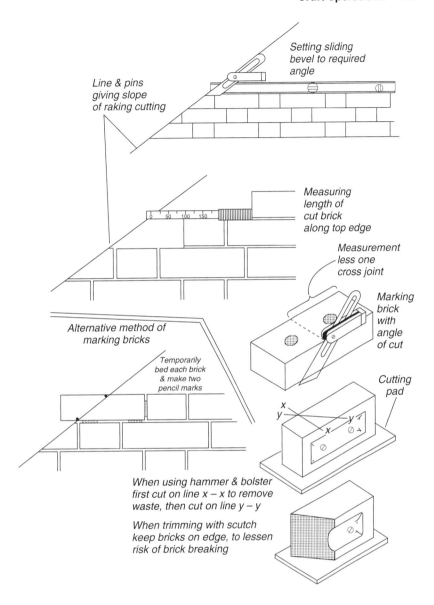

Figure 8.55 Procedure for raking cutting

Alternative method, if sliding bevel is not available

1. Rack back and tooth out as before.
2. Temporarily bed a brick, carefully, to line and level, as shown in Fig. 8.55 at bottom left.
3. Apply two pencil marks on the brick face to indicate the angle of cut.
4. Remove the brick from its temporary bed, and cut it with hammer and bolster; trim with scutch; and permanently bed the cut brick.

If perforated wirecut facing bricks or engineering bricks are being used, these will have to be cut on a masonry bench saw. Mark five or six at a time, indicating clearly the 'waste' part of each brick to be cut. Number each one for easy identification, and send them for cutting on the site masonry bench saw.

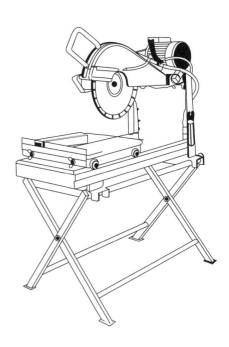

*DO NOT CHANGE
GRINDING WHEEL
UNLESS AUTHORISED
TO DO SO*

Figure 8.56 Masonry bench saw

Safety

Warning. It is an offence under the Health and Safety at Work Act to use a masonry bench saw if you are under the age of 18. Even aged 18 and over, the Act says you must have been instructed in its safe use. So follow these simple but vitally important rules:

Wear eye and ear protection
Ensure that you are wearing no loose clothing
The blade must be secured and fixed by a qualified person
Check that the blade guard is in place
Clamp the brick to be cut firmly on the trolley, and keep your fingers well
    away from the blade
Make sure that the water tank is filled and the pump working
Do not force the blade as it cuts through the brick

Remember

It is the employers' responsibility to ensure you are correctly trained to use power tools. The item of equipment provided must be maintained and safe. Employees in turn have a duty to inform their employer of any work situation which presents a risk to themselves or their workmates.

Brick corbels (Figs 8.57, 8.58 and 8.59)

The building of oversailing courses forming corbels – in 56 mm oversailers for the laying and in 28 mm for bond arrangement – is not an easy operation to perform. In the case of the 56 mm corbels, headers should be used wherever possible; if the use of stretchers cannot be avoided, they

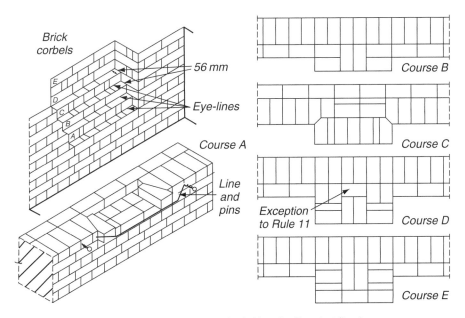

Each corbel course must be laid to the line, but fixed
to the bottom arris, as this forms the 'sight-line'

Figure 8.57 Setting corbel courses to a line

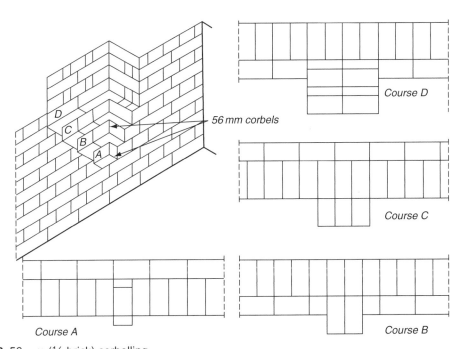

Figure 8.58 56 mm (¼ brick) corbelling

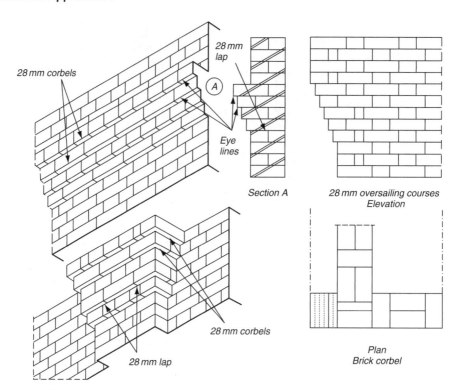

Figure 8.59 28 mm (¹/₈ brick) corbelling

should be the last bricks laid on that course and should be well bedded, a cross-joint being placed throughout the length of the brick.

In the case of 28 mm corbels, the greatest possible lap must be maintained; if special attention is given to this point, and common sense applied, no difficulty should be experienced in carrying out this work.

Both internal and external straight joints are sometimes unavoidable, but they should be reduced to a minimum and the specified bond should be adhered to, if possible. The final shape of the brickwork must be considered; this will continue to a greater height than the corbel and must be practical and not involve unnecessary cutting.

When a corbel is being built it must be kept well tailed down to prevent overturning, that is, the back bricks must be laid first in order to bond the previous corbel before attempting to lay the next oversailing course.

'Soldier' course (Figs 8.60 and 8.61)

Consists of bricks laid on end continuously.

Care should be taken to avoid laying the course at an angle and a small boat-level should be used for checking the work; see Fig. 8.62.

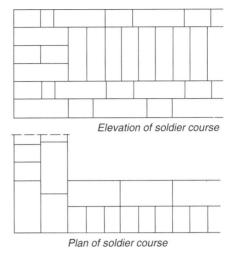

Elevation of soldier course

Plan of soldier course

Figure 8.60 Constructing soldier string course

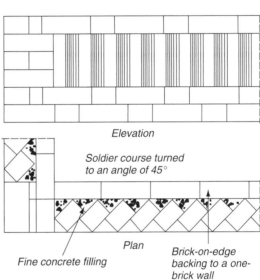

Elevation

Soldier course turned to an angle of 45°

Plan

Fine concrete filling

Brick-on-edge backing to a one-brick wall

Figure 8.61 Soldier bricks turned to angle of 45°

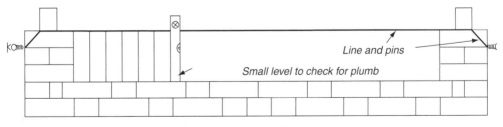

Line and pins

Small level to check for plumb

Craft operation for soldier course

Figure 8.62 Checking soldier course bricks

Dog toothing (Fig. 8.63)

The correct angle must be maintained throughout and alignment must be kept. 'A' shows the treatment in the centre of a wall, the work having been carried out from both ends.

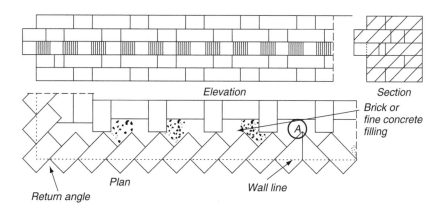

Figure 8.63 Creation of a dog toothing course

Dentil course (Fig. 8.64)

No explanation is needed, except that the eye or sight lines must be maintained.

String or band courses are a decorative feature and are quite separate from normal bonding arrangements. Straight joints between main brick work and the string course must be avoided wherever possible.

Many other designs are possible for string courses by the introduction of basket weave and herring-bone bonds, tiles or bullnose and cant special bricks set as soldiers. All these are laid in the manner already described, eye lines must be maintained and bonding knowledge applied.

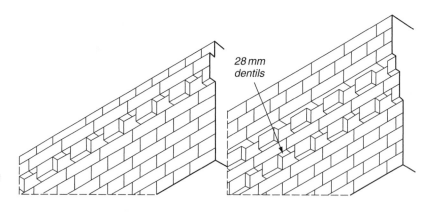

Figure 8.64 Single and double dentil courses

Tumbling-in (Fig. 8.65)

This is a method of reducing the size of a pier or external chimney breast. It is decorative in character and acts as a weathering as shown.

Figure 8.66 shows the preparation, building, and application of buttress tumbling-in for an external chimney, 50 mm × 25 mm batten is a temporary fixing for line. The overhang at 'A' forms a 'drip', thus preventing rain from running down the tumbling directly on to the wall face, and it avoids the ugly joint which would occur if no overhang were introduced. The illustration shows a diaper pattern introduced into a chimney breast as a decorative feature.

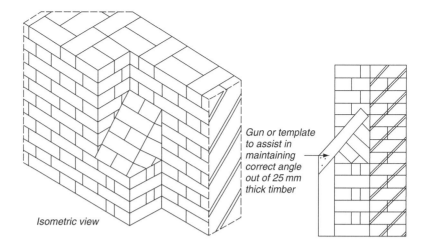

Figure 8.65 Small section of tumbling-in using a gun template

Isometric view

Gun or template to assist in maintaining correct angle out of 25 mm thick timber

Bricks circular on elevation

Curved work on elevation is mainly used in arch construction and will be explained later. Other uses of curved work on elevation will be required on boundary walls and gable ends. The curves can be either convex or concave and are set out similar to curved work on plan and are mainly built with a trammel.

Concave

The brickwork has to be completed to the height of the striking point to support a timber beam. The timber beam is placed on the erected wall and bricks are used to provide weight to hold it in place.

A trammel is fixed to the required striking point in such a way as to allow it to swing easily. The bricks are marked and cut to the shape of the curve formed by the trammel. The trammel can then be adapted to form the curve of the brick on edge. It is usually shortened by the depth of the brick on edge plus a joint. The bricks can then be laid using the trammel to check for accuracy.

If the radius is small the bricks may have to be cut wedge shape to produce an acceptable appearance.

On larger radii, V-shaped joints may be satisfactory.

The circular ramp in Fig. 8.67 shows a brick-on-edge finish, but a brick laid flat may be used, according to the architect's design.

Figure 8.68 shows the method of maintaining the stability of the circular ramp by building in 200 mm long stainless steel fish-tail cramps. Brick-on-edge finish, whether circular or straight, can be strengthened in this manner.

Convex

The problem with convex curves is that the striking point is always set out in the brickwork below the curve.

The trammel has to be secured to the brickwork mortar joints where possible or by first securing a piece of timber to the brickwork (Fig. 8.69).

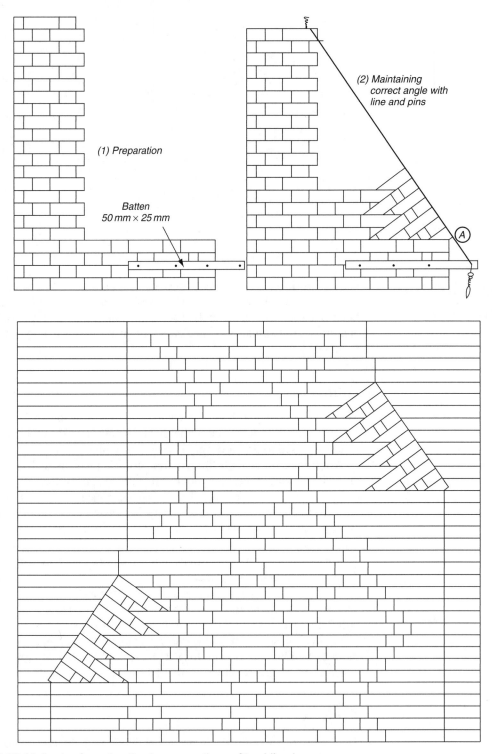

(1) Preparation

*(2) Maintaining
correct angle with
line and pins*

*Batten
50 mm × 25 mm*

Ⓐ

Figure 8.66 Methods of constructing longer sections of tumbling-in

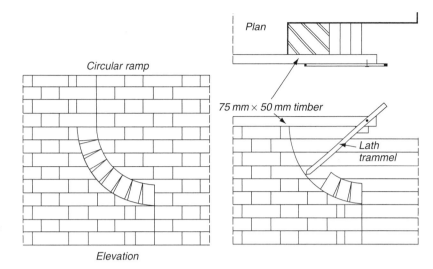

Circular ramp

Plan

75 mm × 50 mm timber

Lath
trammel

Figure 8.67 Use of trammel to construct a concave circular ramp

Elevation

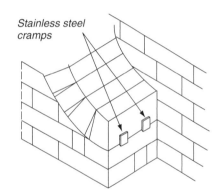

Stainless steel cramps

Figure 8.68 Stainless steel cramps securing first ramp bricks

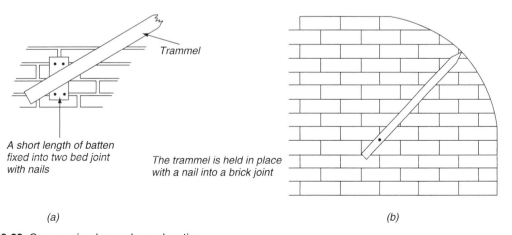

Trammel

A short length of batten fixed into two bed joint with nails

The trammel is held in place with a nail into a brick joint

(a)

(b)

Figure 8.69 Convex circular work on elevation

The trammel is then fixed to the backing timber in such a way as to allow it to swing easily.

Walls curved on plan

Curved brickwork is constructed in modern buildings to provide semi-circular, segmental and elliptical curves. To build this type of work vertical plumbing and horizontal levelling of the courses is most important as it is not possible to line in the brickwork with lines and pins.

To construct curved brickwork *Three* methods are available:

1. Full templates or moulds
2. Small template and plumbing points around the curve
3. Trammel.

Full templates

Templates can be produced when constructing bay windows and are therefore known as bay moulds in parts of the country; see Figs 8.23 and 8.24. They are constructed to give the line of the main wall and also the full curvature of the outer face of the bay brickwork. The outer face of the timber being carefully shaped to the required curve. Such templates are used for semi-circular, segmental and other bay windows incorporating curved brickwork (Fig. 8.70).

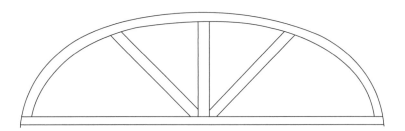

Figure 8.70 Typical template for a segmental bay window

Small template and plumbing points around the curve

If the curve is small it is possible to produce a template to fit the whole curve (Fig. 8.71). When working with larger curves, a small template can be produced approximately 600–1200 mm long and one face shaped to give the exact curvature of the brickwork. The first course of brickwork is laid from a large template or trammel.

A number of plumbing points are marked around the curve.

As each course of brickwork is commenced a brick is bedded in bond at each of these points. These bricks are levelled horizontally from the main wall. Each marked point is plumbed vertically from the first course. The brickwork between these plumbing points is built in using the small template approximately 1200 mm long to check the accuracy of the curve.

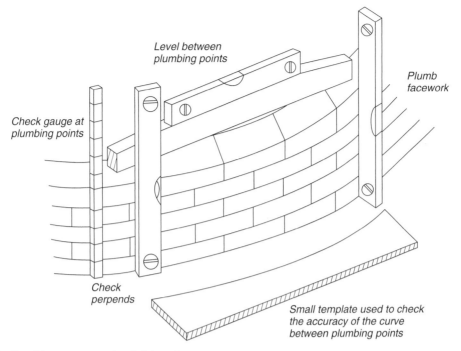

Level between plumbing points

Plumb facework

Check gauge at plumbing points

Check perpends

Small template used to check the accuracy of the curve between plumbing points

Figure 8.71 Checking curve on plan brickwork

Trammels

Curved work can also be set out using trammels as shown in Fig. 8.72. The method used is to set up in a correct position a trammel point. This is

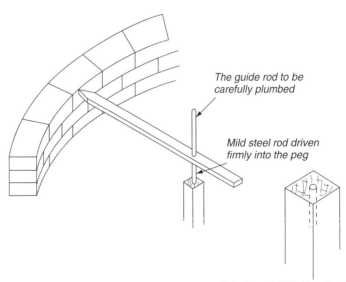

The guide rod to be carefully plumbed

Mild steel rod driven firmly into the peg

Hole for rod drilled into the top of the peg before it is driven into the ground

Figure 8.72 Using a trammel

a rod of mild steel, 20–25 mm diameter set in a wooden peg or concrete block. This pivot when fixed must be vertically plumb.

The trammel is a piece of timber approximately 20 mm × 125 mm. In one end of the trammel a hole is drilled so that it can be placed over the mild steel rod and at the other end a point is cut.

The courses of brickwork are laid horizontally level to the curve given by the swing of the trammel. The trammel can be used for either face of the wall.

Bonding

Header bond

One type of bond which is often used on curved walling is header bond. This bond consists entirely of headers on the face (Fig. 8.73).

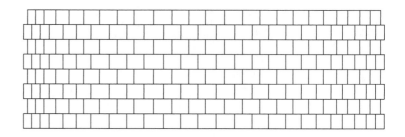

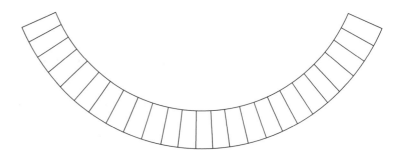

Figure 8.73 Header bond

If the radius is small then alternate methods have to be adopted to avoid large wedge-shaped joints. Only one in three headers traverses the wall, the remainder are snap headers with cut bricks built into the rear of the wall (Fig. 8.74).

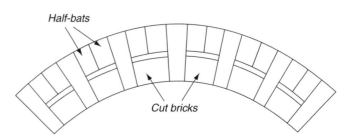

Figure 8.74 Using snap headers on smaller radii curves

Walls that are curved on plan can be built from straight bricks by forming wedge-shaped cross joints (Fig. 8.75).

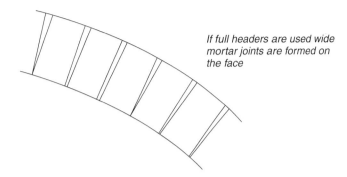

If full headers are used wide mortar joints are formed on the face

Figure 8.75 Wide mortar joints

Stretcher bond may be used in half brick walls to as little as 3 m radius without the need to cut back corners and to give an acceptable cross joint on the face.

A certain amount of cutting may be required on the inside of the wall.

If the facework is required to be neat on both faces it may be necessary to use purpose-made bricks as shown in Fig. 8.76.

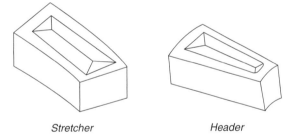

Stretcher   *Header*

Figure 8.76 Specials for curved work

Serpentine walling

This type of walling curves in and out along its length. It is usually confined to boundary walling and gives a pleasant unmonotonous effect. It also may be seen on some large housing estates where the roads are deliberately constructed with curves in them to control the speed of the traffic.

When building walls which are curved on plan, it is important to set out the plumbing points at the base and to maintain these points all the way up the wall.

The work in between the plumbing points should be checked by the use of a template cut to the shape of the curve, out of plywood or hardboard (Fig. 8.77).

Another method for checking the accuracy of curved work with a small radius is to use a radius rod. First, a piece of steel rod is fixed into position and plumbed. A batten is then drilled so that it fits easily over the rod

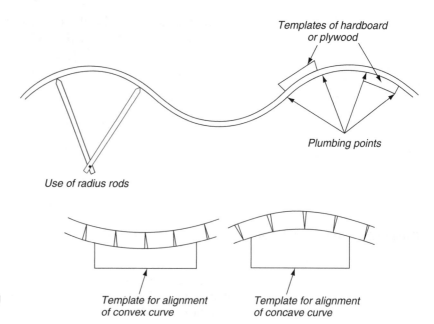

Templates of hardboard
or plywood

Plumbing points

Use of radius rods

Figure 8.77 Serpentine walling
with templates or radius rods

Template for alignment
of convex curve

Template for alignment
of concave curve

and is cut to the length of the radius. The batten is threaded over the rod
and the wall can now be built to the batten. This method is particularly
useful in the construction of walls for spiral staircases and in other similar
types of work carried out in confined spaces.

9 Bridging openings

All buildings have openings of some description in order to allow access and egress and admit light and ventilation. Whenever an opening is made in brick- or blockwork walls a weakness could occur in the walls if the work is not carried out correctly.

Today modern buildings include larger openings than before so it is essential to take extra care when forming the opening and selecting the most appropriate method of bridging the opening to ensure the maximum amount of stability and strength in the wall is maintained.

These openings have to be bridged to support the walling above in a way that will not cause any cracking to the structure either above or below the opening.

There are several methods available but the simplest method is by using a lintel. The lintel carries the weight of the loads above the opening and distributes this load to the abutments either side.

Early civilisations tried various ways of supporting brick and stone masonry across openings in their buildings as shown in Figs 9.1 and 9.2.

Pyramid-building Egyptians and ancient Greeks used lots of columns and stone lintels, but it was the later Roman builders who developed the idea of forming arches from separate blocks of stone, bricks or tiles mortared together.

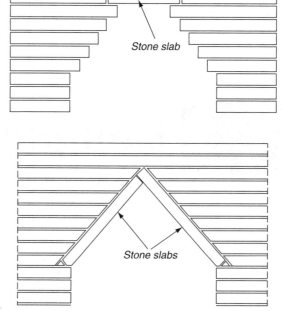

Figure 9.1 Spanning an opening by corbelling

Figure 9.2 Alternative early method of spanning an opening

Properly bonded brickwork is able to 'bridge' itself across openings in a series of offsets as shown in Fig. 9.3. This is called 'natural bracketing' and means that, in theory, only that area of brickwork within the 60 degree triangular shape in Fig. 9.3 needs supporting by a lintel. The greater the span of an opening, the larger this triangular piece will be.

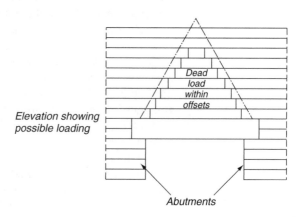

Figure 9.3 Natural bracketing of bonded brickwork above a lintel

Definitions

Lintels

These are straight beams of concrete or steel which are made strong enough to support the dead load of bricks or blocks above. Because lintels are straight, they have a slight tendency to bend or deflect when bricks or blocks are bedded on top. This loading results in tensile stress in the lower part of a lintel, plus compressive stress in the upper part, as indicated in Fig. 9.4.

Steel rods are cast into the lower portion of concrete lintels to absorb this tension, so that the lintel does not crack on the underside when loaded. Concrete in the upper part of the lintel is very good at resisting compressive stress.

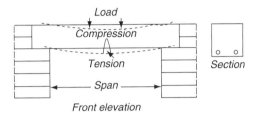

Figure 9.4 Typical stresses in a lintel

Types of reinforced concrete lintel

Concrete lintels can be made available in any one of five ways, depending upon the following site considerations:

- How large do lintels need to be?
- How much will each lintel weigh (mass in kg)?
- How are they to be moved around site and to upper floors?
- Are they to be lifted into place by hand or mechanically?
- Will they be seen in the finished work, or plastered?

Pre-cast lintels

These are lintels that are made in a mould at ground level, and when the concrete has hardened for at least a week, are lifted and bedded in place.

Figure 9.5 shows a typical site casting platform, set perfectly level both ways. Bricks or blocks are used as spacers, in order to obtain lintels of exactly the right width.

Inside surfaces of these lintel 'boxes' should be lightly coated with mould oil to make dismantling or 'striking' easier. When filling the mould a 25–30 mm layer of wet concrete should be spread over the bottom first before placing reinforcing steel bars. The remainder of the concrete should be added and thoroughly compacted by manual or mechanical vibration. The upper surfaces should be trowelled smooth and marked clearly as the 'Top' surface, so that the steel bars will be towards the soffit when permanently bedded (Fig. 9.6).

Factory made reinforced concrete lintels

For larger requirements pre-cast reinforced concrete lintels can be ordered from a specialist supplier to suit the range of openings in a building. Care should be taken to see that a 'fair-face' is specified where lintels will be

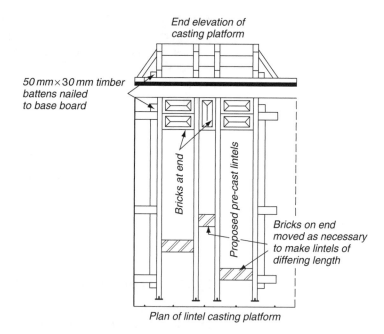

End elevation of casting platform

50 mm × 30 mm timber battens nailed to base board

Bricks at end

Proposed pre-cast lintels

Bricks on end moved as necessary to make lintels of differing length

Figure 9.5 Method of casting lintels on site

Plan of lintel casting platform

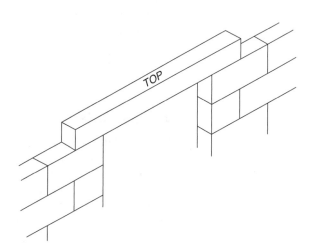

TOP

Figure 9.6 Pre-cast reinforced concrete lintel in position

permanently exposed, and that the 'top' surface of all lintels delivered to site is clearly marked (Fig. 9.6).

Pre-stressed concrete plank lintels

These are a special type of reinforced concrete lintel which can only be made in a factory.

The high tensile steel 'strands' are stretched by hydraulic jacks before the concrete is compacted in the lintel mould. When the concrete has thoroughly hardened, the stretched steel reinforcement is carefully released and keeps the whole lintel section in a permanent state of compression. These lintels are known as 'plank' lintels because they are only 65 mm deep for all spans up to 1800 mm (Fig. 9.7).

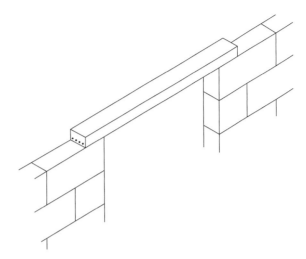

Figure 9.7 Pre-stressed concrete plank lintel in position

Cast in-situ concrete lintels

This is where a lintel would be much too heavy to lift by hand if pre-cast at ground level. Instead, a timber box or mould is constructed exactly where the lintel is required. This formwork, Fig. 9.8, is made

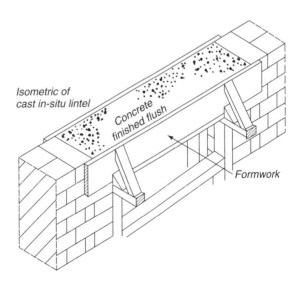

Figure 9.8 Formwork for a cast in-situ lintel

strong enough to withstand the pressures of filling with wet concrete, and thoroughly compacting it.

The concrete is poured around steel reinforcing bars placed approximately 25 mm up from the soffit; see Fig. 9.9.

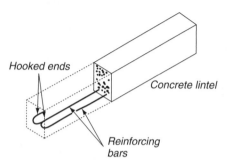

Figure 9.9 Reinforcement in lower part of lintel

Bedding lintels

Lintels must be set upon a wet mortar bed joint at each side, so as to spread the load evenly over the whole bearing surface. When tapping down into position the spirit level should be held against the underside of the lintel in case the top edge is not parallel with the soffit.

Steel lintels

Sheet steel can be pressed into shape in a factory to make lintels that are much lighter to lift than concrete, and which can withstand both tensile and compressive stresses. These are available in several shapes and sizes.

Internal steel lintels

Internal steel lintels are designed mainly for 100 mm walls but other sizes are available, and can be either boxed, corrugated or channelled; see Figs 9.10, 9.11 and 9.12.

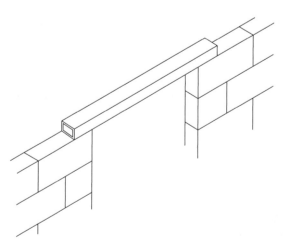

Figure 9.10 Boxed steel internal lintel

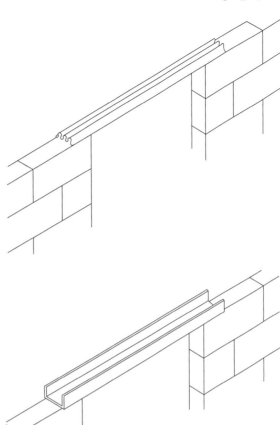

Figure 9.11 Corrugated steel internal lintel

Figure 9.12 Channelled steel internal lintel

External steel lintels

External steel lintels are designed mainly for cavity walls and are available in various designs; see Figs 9.13, 9.14 and 9.15.

Lintel bearings

A solid seating or bearing is required at each end of a lintel to support the concentrated pressure of one half of the total load resting upon the lintel. This bearing at each side should be a minimum of 150 mm for lintel spans up to 1800 mm, with 225 mm minimum up to 3 m span. Thereafter, a structural engineer will be required to calculate the safe bearing area needed for larger span lintels. BS 5628:1985 Part 3 'Workmanship' recommends that this bearing level should coincide with a whole block rather than a half or cut block, for greater stability where lintels are set in block walling; see Fig. 9.16 and all previous drawings.

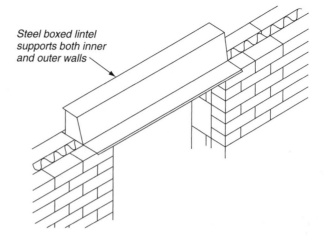

Figure 9.13 Steel section lintel for cavity walls

Steel boxed lintel supports both inner and outer walls

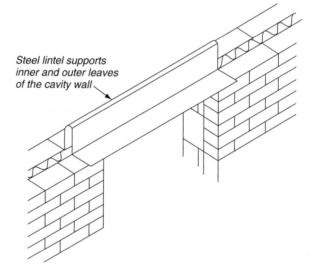

Figure 9.14 Bent steel lintel for cavity walls

Steel lintel supports inner and outer leaves of the cavity wall

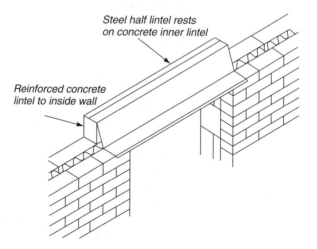

Figure 9.15 Steel half lintel for cavity walls

Steel half lintel rests on concrete inner lintel

Reinforced concrete lintel to inside wall

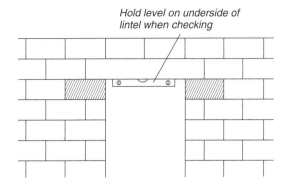

Figure 9.16 Whole block under lintel bearings to spread the load

Reinforced brick lintel

The bricklayer calls this a 'soldier arch', as the bricks are placed on end resembling a file of soldiers.

Care should be taken to avoid laying the bricks at an angle and a small boat-level should be used for checking the work, as explained previously with soldier courses; see Fig. 9.17.

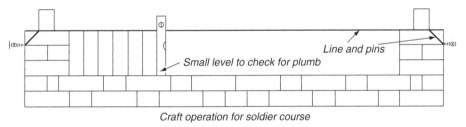

Craft operation for soldier course

Figure 9.17 Checking soldier course bricks

Before setting the lintel, temporary supports are placed (see Fig. 9.18). Several methods of reinforcing brick lintels are shown (see Fig. 9.18); at 'A' where a 225 mm soffit occurs, steel bar reinforcement is placed throughout the length of the lintel, with stainless steel ties positioned at intervals of approximately 225 mm; at 'B' the brick lintel is erected with wire reinforcement built in and left projecting, to allow the concrete lintel to be cast in situ around it, thus giving adequate stability.

This type of lintel, if erected truly horizontal and plumb, causes an optical illusion to occur at a distance of 3 metres or more from it; it appears to sag. This appearance can be overcome by giving the lintel a slight camber or by giving a curvature to the soffit along its length, together with a slight skewback. The adjustment must be made carefully and with the consent of the architect.

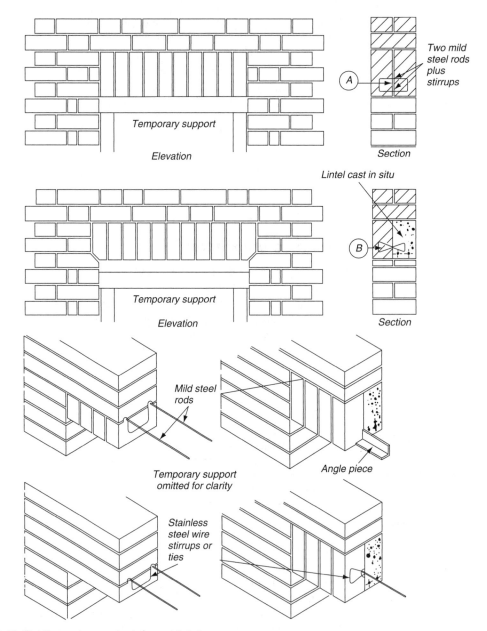

Figure 9.18 Soldier arches and reinforced lintels

Brick arches

Arch construction is a more decorative means of spanning openings. True arches are made to curve upwards when looked at in elevation, so that they are always in a state of compression, wedged between abutments. As bending under load cannot take place, there will be no tensile stress in the arch.

All arches are formed upon a temporary support called an arch 'centre' which must not be removed until the jointing mortar has been allowed to harden for at least a week. Only when the arch centre has been removed does the arch become self supporting.

Classification

Arches

The names of the various parts of an arch are shown in Fig. 9.19. Arches are named or classified in the following ways:

1. The method of cutting or fixing, e.g. rough ringed, axed, gauged
2. Their shape, e.g. Gothic, camber or flat, semi-circular, segmental
3. The number of centres from which they are set out.

These classifications can be used in combination, e.g. axed segmental, gauged three-centred semi-elliptical.

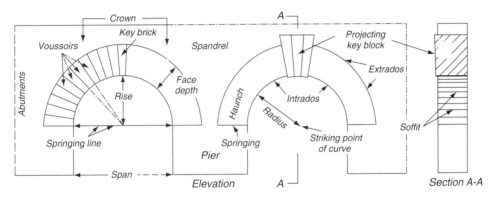

Figure 9.19 Craft terms in arch construction

Rough ringed arch

This type of arch is the simplest to construct. It is formed from uncut bricks and its shape is controlled by the type of turning piece or centre adopted. The joints are wedge shaped and thus play an important part in the stability of the arch. In the axed or gauged arch the reverse applies – bricks are wedge shaped and the joints are parallel.

The arch is constructed of a number of half-brick rings, hence the name ringed arch, the number of half-bricks varying according to the size of the opening. This method is adopted to reduce the size of the mortar joint at the extrados of each separate ring (Fig. 9.20). In the illustration, the use of stretchers to turn the arch would make the width of the joint at the extrados excessive.

Over large spans, and where the arch is several rings in depth, lacing courses (three or four courses of axed work) can be inserted at intervals throughout the depth and thickness of the arch as a decorative

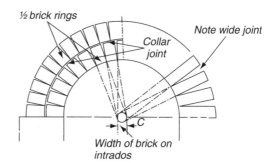

Figure 9.20 The reason for using two header rings and not stretchers for rough arches

feature (Fig. 9.22). Figure 9.21 shows a $1^1/_2$ brick arch made up of two rings, with the introduction of lacing courses at skewback and key. It will be noted that the bottom ring is one brick in depth and bonded; this is possible where the arch is of a wide span and of easy curvature.

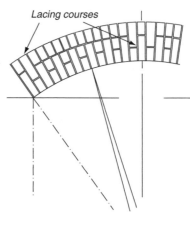

Figure 9.21 Elevation of a rough ringed arch

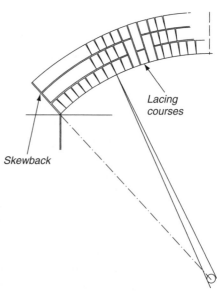

Figure 9.22 Alternative face bonding to Fig. 9.21

In the days when walls were built of flints, horizontal bands of brick-work, three or four courses deep, were introduced at various intervals. These were called lacing courses, and used in this way they helped to strengthen the wall structure.

Relieving or discharging arch

This is a form of ringed arch, built over a lintel, and turned on a shaped brick core (Fig. 9.23). It relieves the lintel of any point load and discharges

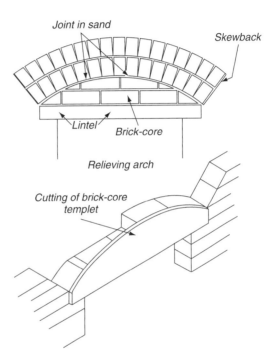

Figure 9.23 Formation of relieving or discharging arch

the weight to the abutments. The joint between the brick core and the soffit of the arch must be of sand, as this ensures the proper discharging of loads. This sand joint is pointed on the face only. The shape of the brick core is obtained by the use of a templet.

Axed arch

This type of arch is formed by bricks cut to wedge shape from the ordinary facing brick. It can be of the same colour and texture as the wall facings or of some contrasting colour and probably of finer texture.

The tools used for cutting and trimming are the hammer, bolster and scutch, with a carborundum block.

The thickness of the mortar joint varies from 4 mm up to 10 mm.

The operations of cutting and setting will be explained later in this chapter.

Brick arches – gauged work

Earlier editions of this book contained a whole chapter describing the bricklayers' craft operation of cutting and rubbing gauged arch brickwork. This work requires specially made clay bricks called 'red rubbers'. These bricks contained approximately 30% fine sand mixed with the clay, and were carefully fired so that the same even orange-red colour and texture was present throughout the body of each brick.

The high content of sand allowed these bricks to be hand sawn and rubbed, on a block of York stone, to permit joints of 1 mm thickness. Gauged work demanded the very highest skills from the bricklayer in setting out, cutting, rubbing and bedding these bricks, which were 'white-line' jointed, using a putty made only from freshly slaked lime and water.

Much thought was given before deciding to omit this chapter (which fully described this highly skilled aspect of the bricklayer's work) from this revised edition. The primary reason for leaving it out is that these red rubber bricks are no longer available; another reason is that modern methods of cutting voussoirs on masonry bench saws have displaced the labour-intensive traditional method of cutting and rubbing by hand.

Masonry bench saws can cut plain red facing bricks to shape without the high sand content, and still obtain the fine 'white-line' joints that distinguish the appearance of gauged work from other face brickwork.

Types of arch

This section will deal with the shape of arches and their geometrical setting out. These arches will all be shown straight in plan. Various simple geometrical constructions are illustrated and explained to assist in the understanding of arch setting out and correct construction on site; see Figs 9.24 and 9.25.

1. *To bisect a given line AB*. With the compasses set at a radius greater than half the length of the line, and used successively at points A and B, describe the intersecting arcs AC, AC' and BC, BC'. Then CC' is the perpendicular bisector.
2. *To bisect the angle ABC*. Set the compasses at any radius and from point B describe arc DD'. With centres at D, D' alternately, and with the same radius or a radius greater than a half DD', describe the intersecting arcs E. Then BE is the bisector.
3. *To erect a perpendicular on line AB from point A*. From any centre C and with the distance CA as radius, describe a circle cutting AB at D. Draw the line DCE. Then the line AE is the perpendicular.
4. *To divide a given line into a given number of equal parts*. From one end of the line AB, draw a straight line at any angle. From point A on this line mark off the given number of equal lengths, say five. Join B5. Through the other points draw parallels to the line B5. Then the line has been divided into five equal parts.
5. *To draw a tangent through a given point C on the circumference of a circle*. Draw the line AB from the centre of the circle to cut the

circumference at C, the given point. Make CB equal to AC and bisect AB by drawing arcs D and E as in Method 1. Then line DE is at a tangent to the circle. Note that the intrados and extrados of an arch are made up of a series of tangents through its voussoirs. A line taken through the centre of a voussoir from the striking point is at right angles to this tangent (Fig. 9.24).

6. *To draw a circle to pass through three given points A, B and C.* Join AB, BC. Bisect the lines or chords AB, BC. Then the centre of the required circle, point D, is at the point where the perpendicular bisectors intersect (Fig. 9.24).

7. Figure 9.25 illustrates circles in external and internal contact. They touch one another and are said to be at a tangent. Notice that their centres are on the same line AB, which passes through the point of contact and that a tangent drawn at C is common to both circles. Line AB is called the 'common normal'.

8. An application of 7 is that circles in internal contact form the curve of a three-centred or semi-elliptical arch. Note that a three-centred arch is not truly elliptical; it follows a similar shape and is therefore called elliptical, but if a true ellipse were formed, each brick would need a

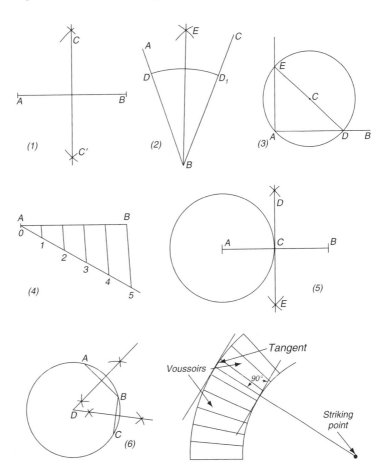

Figure 9.24 Geometry of arches

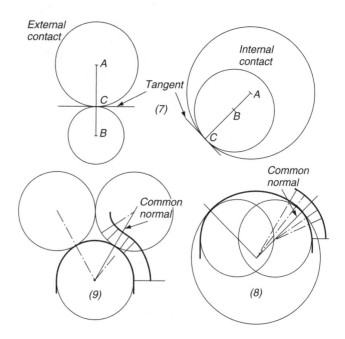

Figure 9.25 Geometry of arches

separate templet. By forming an approximate ellipse of three or five centres only, two or three templets respectively are needed. This will be further explained in a later section.

9. A further application of 7, showing circles in external contact forming an ogee curve.

Segmental arch

Figure 9.26 shows the geometrical setting out of a segmental arch. The rise to any segmental arch is normally one-sixth of the span. First draw the span, assume it to be 900 mm, and set up a perpendicular bisector. Mark off the rise, 150 mm. Join AB and bisect. Where this bisector intersects the centre line, point C, is the striking point of the required arch. A face depth of 225 mm is shown.

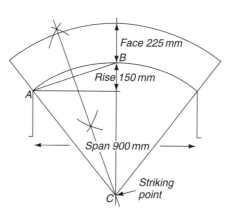

Figure 9.26 Drawing of segmental arch with rise of one-sixth span

Figures 9.27 and 9.28 show the elevation of ringed and axed arches. To ascertain the number of voussoirs and the position of the joint lines in the axed arch, set the dividers at 75 mm, place the points equidistant on each side of the centre line, thus forming the key brick and step round the

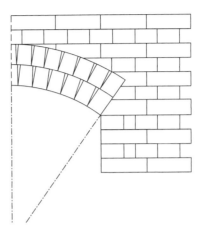

Figure 9.27 Two-ring rough segmental arch

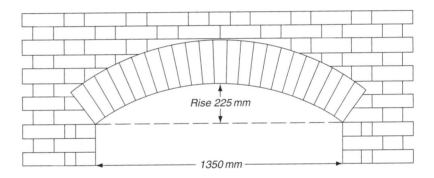

Rise 225 mm

1350 mm

Figure 9.28 Axed segmental arch

extrados. If the last step fails to connect with the springing point at the first attempt, make a further attempt by slightly closing the dividers. The equivalent size, 75 mm, is accepted in general drawing practice for both axed and gauged arches. For the full-size setting out, however, in preparation for the cutting of an axed arch, the voussoir size is taken from the type of brick being used.

There is no need in practice to draw or set out a ringed arch, but for drawing to scale, as a means of illustration, draw the intrados and extrados, step out the brick size on the intrados and obtain the wedge-shaped joints by describing a circle equal to the size of brick, at centre C (see Fig. 9.20).

Semi-circular arch

Figure 9.29 is self-explanatory. Note the joints radiating from the centre or striking point. One face templet is needed for the cutting of this arch.

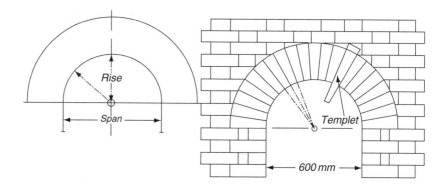

Figure 9.29 Construction of a semi-circular arch

Gothic arch

All arches under this heading are of the pointed type, and each has its particular name. Observe that no key brick occurs.

Equilateral (Fig. 9.30)

With point A as centre and AB as radius, describe an arc from the springing line to the centre line. The intrados and extrados are concentric (i.e. have the same centre). Repeat movements on the opposite springing, with point B as centre. This is a two-centred arch. Complete the setting out of the voussoirs

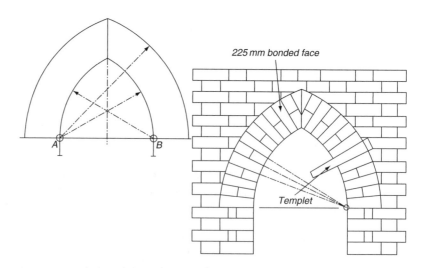

Figure 9.30 Construction of an equilateral Gothic arch

as in the segmental arch, except that there is no key brick. One face templet is needed.

Drop Gothic (Fig. 9.31)

Draw the springing line and assume a span of 600 mm. Set up the centre line and mark on this the rise; assume this to be 400 mm. Draw the chord AB and bisect. The point where the bisector cuts the springing is the striking point of

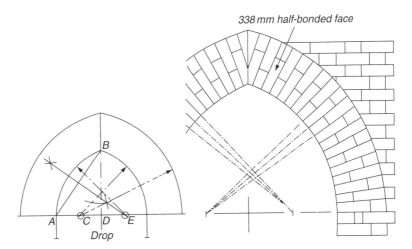

Figure 9.31 Construction of a drop Gothic arch

the arc required. To obtain the opposite striking point, make DC equal DE. The striking points are inside the springing points.

Lancet (Fig. 9.32)

Draw the springing line and extend it beyond the reveal lines. Mark the span 600 mm and the rise 675 mm, and proceed as in the drop arch. The striking points fall outside the springing points of the arch.

The position of the striking points denotes the type of arch. Thus, for 'drop' or 'lancet' arches the striking point is inside or outside the reveal

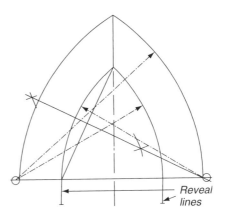

Figure 9.32 Construction of a lancet Gothic arch

lines respectively, while for the 'equilateral' arch it is constant, that is, on the springing points.

Semi-Gothic (Fig. 9.33)

This is an arch with a semi-circular intrados and a Gothic extrados. The depth of the face of the arch at the springing is 225 mm and at the crown

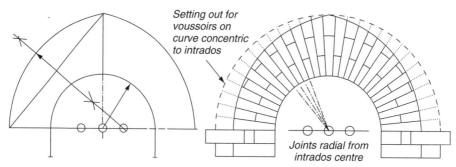

Figure 9.33 Florentine or semi-Gothic arch

338 mm. To obtain the Gothic extrados, follow the method used for the drop or lancet arches. The setting out of the voussoirs on the extrados is taken on a curve which is concentric to the intrados; this regulates the size of the voussoirs. Note the key brick.

Venetian Gothic (Fig. 9.34)

The intrados and extrados are Gothic in shape, but have a different radius, given by a 225 mm face depth at the springing and a 338 mm face depth at the crown. The voussoir sizes are obtained by the method used for the semi-Gothic – a curve concentric to the intrados. Figure

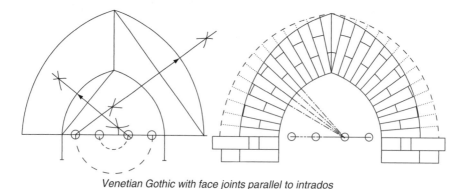

Venetian Gothic with face joints parallel to intrados

Figure 9.34 Venetian Gothic

9.34 illustrates the cross joints of the arch concentric to the intrados, and Fig. 9.35 shows them symmetrical to both intrados and extrados. The method adopted depends on the architect's instructions. To find the centres for the cross joints in Fig. 9.35, divide the face of the arch at the springing into three equal parts and follow the same procedure at the crown. Assume the parts to be complete arcs, construct the chord lines and bisect, following the methods shown for striking the intrados and extrados arcs.

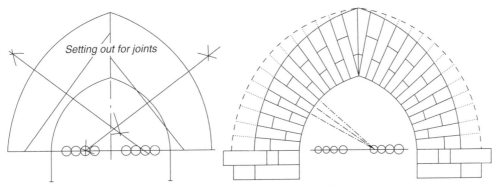

Venetian Gothic with face bonding in proportion from springing to crown

Figure 9.35

Segmental Gothic (Fig. 9.36)

A Gothic arch with a skewback, which probably derives its name from this fact. It is formed by two segments meeting and forming a pointed arch. Set out the springing line and the rise. Join AB and bisect; where the bisector intersects the reveal line the striking point of one half of the arch occurs. Mark a similar striking point on the opposite reveal to complete the setting out.

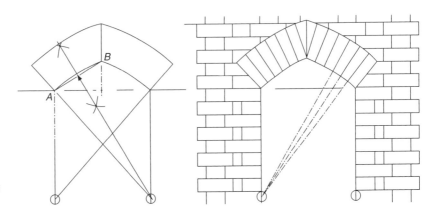

Figure 9.36 Segmental Gothic arch

Elliptical or Tudor Gothic (Fig. 9.37)

This is a four-centred arch, the composite curve being constructed by the application of circles in internal contact. Several methods can be adopted for the geometrical setting out of this arch, but the method illustrated is universal, as it permits of the construction of the arch to a given rise and a

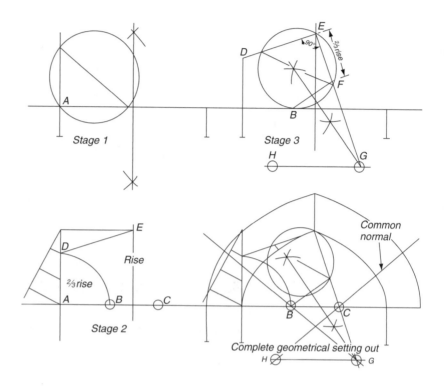

Stage 1

Stage 3

Stage 2

Complete geometrical setting out

Figure 9.37 Tudor arch

given span. Two face templets are needed for its cutting. The setting out is shown in its various stages:

1. Assume the rise to be 450 mm and the span to be 900 mm. Draw the springing line and set out the span. Erect the perpendicular bisector between springing points and set up a perpendicular at A.
2. Mark off the rise from the centre point of span to E, and set out two-thirds on the perpendicular at A. With the point of the compasses at A and with a radius of two-thirds the rise, describe a quadrant to intersect the springing line at B. This is the first striking point. Ascertain point C equidistant from the centre line at B, to make the second striking point. Join DE.
3. Draw a line at 90° to DE at point E and on this mark off two-thirds of the rise EF. Join F and the first striking point B, and bisect. Where this bisector meets the line produced from E, the third striking point G occurs. Ascertain point H equidistant from centre line as G, to make the fourth striking point. Note the common normals through striking points GB, HC. In all arches where different curves are employed, make the common normal a joint line and set out voussoir sizes on the respective curves on each side of this. Thus in Fig. 9.38, X bricks between S and T, and similarly between T and U. Figure 9.39 shows an arch with a face depth of 338 mm bonded quarter-bond and Fig. 9.38 an arch with a 225 mm face and the two templets required for the cutting.

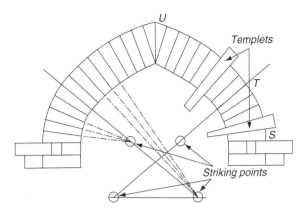

Figure 9.38 Elevation of
Tudor arch

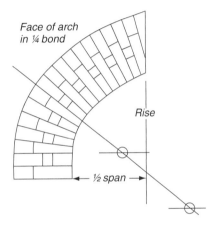

Figure 9.39 Half elevation of
Tudor arch, bonded on face

Semi-elliptical or three-centred arch (Fig. 9.40)

Another example of circles in internal contact. As in the case of the Tudor
arch, several methods are possible in the geometrical setting out, but the
method shown is general. Assume the rise to be 300 mm and the span to
be 900 mm.

1. Set out the springing line and the rise.
2. Join AB. With the point of the compasses at C and with a radius equal
   to half the span, describe a quadrant to intersect the centre line at D.
   With the point of the compasses at B and with a radius equal to BD,
   describe an arc to cut AB at E.
3. Bisect the line AE, and where this bisector intersects the springing line
   and centre line the first and second striking points occur respectively.
   The third striking point G is on the springing line equidistant from the
   centre line, as F.

The figure is self-explanatory.

As previously explained, the three-centred and five-centred arches are
approximate ellipses. The more centres used in the geometrical construction,

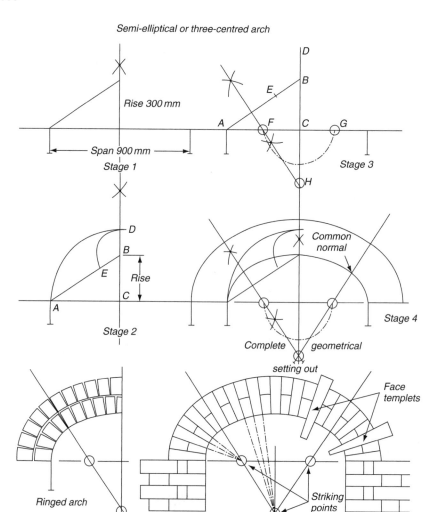

Figure 9.40 Semi-elliptical or three-centred arches

Semi-elliptical or three-centred arch

the truer the ellipses become. For an arch span of up to and including 1350 mm, three centres give a pleasing curve. Beyond this size and up to, say, 3 m – the span limit which a bricklayer meets in general practice – five centres should be used.

Camber or flat arch (Fig. 9.41)

The name of this arch is explained by the fact that the arch is intended to be perfectly horizontal, but in order to prevent the optical illusion of sagging, it is given a slight camber or rise. At one time, the arch was given a very acute skewback, but it was found to fracture across the top points of the skewback.

To set out:

1. Draw the springing line, and parallel to this and 300 mm above it, draw the setting out line.
2. Erect a perpendicular at the springing point to intersect the setting out line. From this intersection, mark off the skewback, allowing 25 mm of skewback to every 300 mm of span. Thus, for a 900 mm span, a 75 mm allowance is given; for a 1350 mm span, 112.5 mm; and so on. This inclination is constant for any depth of arch face (Fig. 9.41).

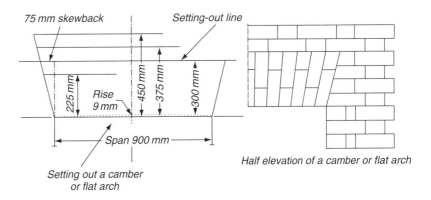

Figure 9.41 Camber or flat arch

The amount of camber given to the soffit is 3 mm to every 300 mm of span, thus for a 900 mm span the rise is 9 mm, and for a 1350 mm span, 14 mm.

In the setting out of the voussoirs in a camber arch, some architects require a stretcher to occur on the soffit at both the key and skewback (Fig. 9.41). To effect this, the number of voussoirs must be a multiple of four, plus one. This procedure is not always adopted, however, as it is sometimes not considered to be economical or practical. Wherever possible, place a stretcher at the key with either a stretcher or header at the skewback, as the case may be.

Moulded segmental arch (Fig. 9.42)

The setting out for the segment is the same as the construction previously explained, but this arch differs from the ordinary segmental arch at the springing point. Where the mouldings on the reveal and soffit of arch intersect, a mitre occurs and the skewback is projected from this, as illustrated. Two methods of obtaining this mitre are possible. Method A (see illustration) is the easier; in this case, a 56 mm moulding is shown. Draw the line where the soffit and reveal mouldings intersect and this is the mitre. Method B is geometrical, and although not always adopted, is a useful one. The setting out is as follows:

Draw the springing line, mark in the rise and set out the intrados curve in the manner previously described. Draw a line to connect the springing and striking points AB. From point B draw a line at 90°, and bisect the

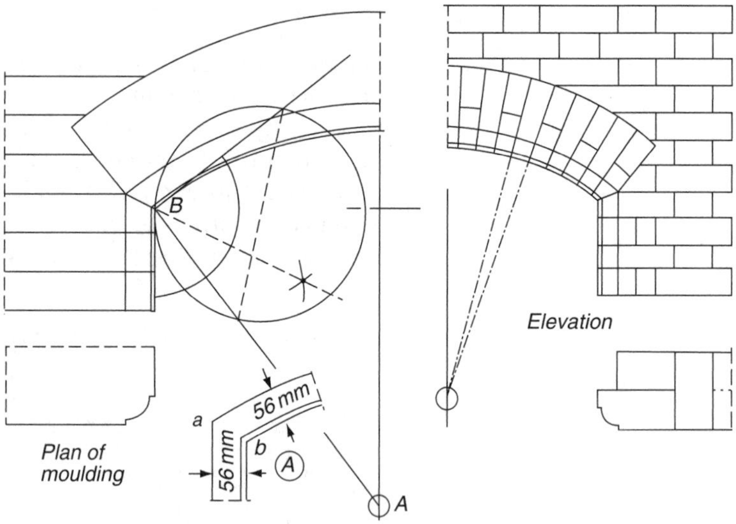

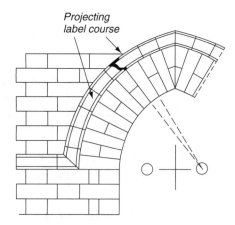

Figure 9.42 Moulded segmental arch

resulting angle between this line and the reveal line; this is the mitre required. Notice the arrangement of skewback voussoir and the moulded reveal brick necessary for proper construction.

Figure 9.43 shows a label course constructed over a drop Gothic arch.

Figure 9.43 Drop Gothic arch with moulded label course

Figure 9.44 shows a 'Welsh arch', a useful method of covering ventilator openings or for rainwater outlets.

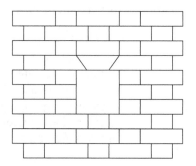

Figure 9.44 Elevation of wall showing a Welsh arch

Cutting an axed arch

Setting out

If an axed arch is to look right, each voussoir must have exactly the same tapering shape, so that mortar joints between them are parallel.

It is now common practice for brick manufacturers to calculate the necessary degree of taper, using CAD (Computer Aided Design) machines, when these arch voussoirs are to be supplied as moulded, specially shaped bricks.

Traditionally, axed brick arches have been produced by cutting bricks using hammer and bolster, after making a pattern templet of the voussoir shape required. This templet is made from a piece of timber 6 mm to 12 mm thick, having first drawn the arch (or half of its elevation) full size on a sheet of plywood.

The craft processes of setting out an axed arch to produce a templet, and then checking it by 'traversing', are dealt with in the following pages.

The tools required are:

1. Trammel heads
2. Dividers
3. Bevel
4. Traversing rules
5. Measuring rule
6. Straight edge
7. Carpenter's tools for cutting wood templet

It is only necessary to set out half the arch, to include the key brick. For the purposes of this description, a segmental arch has been chosen.

Draw the centre line and set off the springing line at right angles to this. Ascertain the springing point and draw intrados, extrados and skewback (Fig. 9.45). Set the dividers to the type of brick being used, and mark out

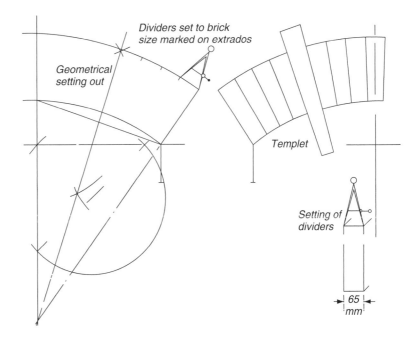

Figure 9.45 Setting out an axed arch

voussoirs on the extrados. Having once set the dividers, do not open them any further, but, if necessary, close them a little. If opened, and a templet is made to the resulting voussoir shape, the bricks will not 'hold out' or will be too narrow to use. Mark out the templet to project approximately 50 mm above extrados and 150 mm below intrados. The templet can be made from a piece of wood 6 mm to 12 mm in thickness, and cut to a taper.

Traversing the arch

Making the templet shape, as described above, gives only the approximate size. To obtain the true shape, it must be traversed or traced over the face of the arch. Traversing is simply a way of multiplying any slight error in the tapered timber shape. To do this, follow Fig. 9.46.

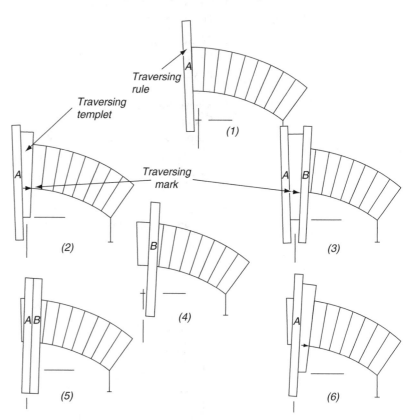

Figure 9.46 Traversing a templet for an axed arch

1. Place the traversing rule A to the key brick.
2. Arrange the templet to fit a voussoir and mark a line on the side of the templet to coincide with the intrados. This is called the 'traversing mark'.
3. Place the traversing rule B.
4. Remove A and the templet.
5. Place A to B.
6. Remove B and again fit templet, allowing traversing mark to coincide with the intrados.

Repeat these operations until the skewback is reached.

If the templet 'fills in' faster at the extrados than at the intrados, i.e. if it touches the top of the skewback but does not reach the bottom, it must be made smaller at the top by being planed down. The reverse applies if the templet fills in faster at the intrados. If, having been traversed, the templet is parallel to the skewback but fails to reach it, or 'underfills' the area between the key and the skewback, the templet must be brought down by placing the traversing mark higher up the templet. If the templet 'overfills', then the traversing mark must be lowered. The operation of traversing is repeated until the templet fits exactly between the key and the skewback, having in mind the number of voussoirs first required in the setting out drawing.

Traversing is an important operation in the construction of arches.

Having obtained the correct shape of the templet, the voussoirs cut to this templet will find their true place in the arch.

Making the joint allowance

Allowance must be made for the mortar joint, and the cutting mark is obtained as follows (Fig. 9.47):

1. Place the templet between two traversing rules.
2. Mark a line on one traversing rule to coincide with the traversing mark on the templet.

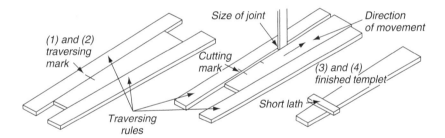

Figure 9.47 Applying the cutting mark to the templet to give joint allowance between voussoirs

3. Holding the traversing rules firmly, tap the templet up until the gap between the traversing rule and the templet gives approximately a 10 mm joint.
4. Transfer the mark on the traversing rule (which was the traversing mark on the templet) to the templet; this is the cutting mark. Ascertain the bevel (Fig. 9.48), remembering that the bevel is at a tangent to the curve of the arch and is found by taking the mean of the squares drawn from both sides of the templet. Fix a short length of lath to the templet at this bevel and the templet is ready for use.

Cutting arch voussoirs

Consideration must be given to the choice of facing bricks for axed arches. Bricks should only be selected that can be conveniently cut with a hammer and bolster into a tapering shape, without excessive wastage.

Figure 9.48 Squaring from both sides of the templet

Application of templet

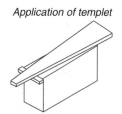

Figure 9.49 Using templet to mark voussoirs

Joggle

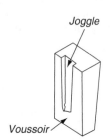

Voussoir

Figure 9.50 Joggling voussoirs with comb hammer

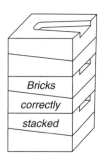

Bricks correctly stacked

Figure 9.51 Voussoirs stacked correctly

Mechanical masonry bench saws can be used for this cutting work with harder or perforated bricks. Class A and B engineering quality bricks will be very difficult to cut by any method, and are really unsuited to axed arch construction. If an arch of engineering quality Class A or B bricks is needed, they are best produced by the brick manufacturer as moulded or special shaped bricks.

Cutting

The tools required are:

1. Hammer
2. Bolster
3. Pencil
4. Scutch
5. Comb hammer
6. Templet
7. A piece of carborundum, to be used to give a clean, sharp arris to the voussoirs.

Apply the templet (Fig. 9.49), scribe with a pencil, and proceed with the cutting. When finished to the required shape, cut a joggle (Fig. 9.50), which allows the arch to be grouted when set in its position, giving added security. (Grout is a mixture of neat cement and water made into a thin slurry.)

Stack the bricks carefully when cut (Fig. 9.51). A well-cut arch should stack evenly, each brick fitting snugly to the one above and below it.

Cutting an axed camber arch

The cutting of the camber arch is different from any other arch in that the bevels are not at a tangent to the same curve and are therefore all different. Some bricklayers consider this arch to be the most difficult to prepare and cut, but if the operations are carried out systematically, no difficulty should arise.

Preparing the templet

Set out the arch as previously described; only half the arch is necessary for practical purposes. There are two methods of drawing the camber:

1. Insert a small tack at the springing points and the top of rise respectively. Spring a lath (Fig. 9.52) arch between these points and draw a pencil line round it. This method gives a suitable curve, and although it cannot be used on an arch with a deep rise, it is quite adequate for a camber arch.
2. The camber slip method (Fig. 9.53) gives a true curve and is suitable for an arch with any rise. Construct a triangular framing of laths, insert a tack at each of the springing points, and place the framing to these

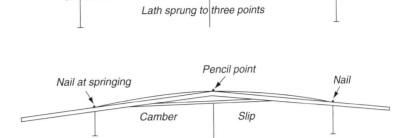

Figure 9.52 Single lath to draw camber soffit

Figure 9.53 Three laths tacked together to draw camber soffit

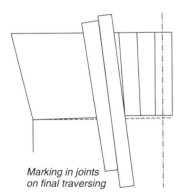

Marking in joints
on final traversing

Figure 9.54 Drawing joint lines

tacks. The apex of the triangle must be at a distance equal to the rise, above the springing line. Place a pencil at the apex of the framing and trace it round, always keeping the arms of the triangle in contact with the tacks at the springing points.

Mark the voussoirs on the extrados, drawing in the key brick shape only, for the time being. To obtain the approximate size at the intrados, divide the distance between the key and the skewback into the same number of parts as there are voussoirs in the extrados. Make the approximate sized templet and traverse as previously described. When the correct size of the templet has been finally reached, traverse again, marking in the joint lines while doing so (Fig. 9.54). This will give the precise size of the voussoirs at the intrados, having in mind that each is slightly different. The obtaining of the correct size is important for the purposes of setting or fixing the arch, as will be described later.

An alternative method of obtaining the joint lines is to produce the skewback lines downwards towards the centre line of the arch, thus obtaining a striking point from which to mark the joint lines. This would, however, be at a considerable distance from the springing line and the method shown above is much more convenient.

The final preparation of the templet differs from that required for the ordinary arch. As already stated, each brick has a different bevel, which must be cut on the brick before it is reduced to its voussoir size. It will be noted that the natural face of the brick can be maintained on the soffit of a normal arch, but with the camber arch this is impossible. To complete the preparation of the templet, therefore, transfer all the soffit bevels from the drawing to the templet, numbering from the skewback to the key (see Fig. 9.55). If the arch is bonded, as in a 300 mm face, letter 'H' or 'S', header or stretcher, according to the brick's position on the soffit. All the information needed will then appear on the templet and this is a wise precaution as a drawing may be lost or obliterated. Note that the practice of marking the bevels on the templet is for convenience only. Remember that the templet is wedge-shaped, so that if the bevels are placed on the templet from the right-hand side, they must be taken off in a similar manner.

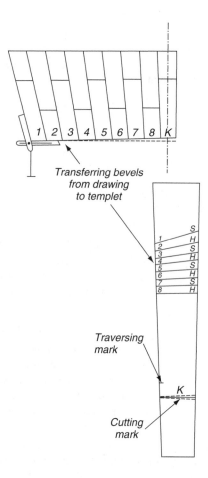

Figure 9.55 Transferring bevels from drawing to templet

Cutting

The camber arch has right- and left-hand sides, again differing from the ordinary arch, where a voussoir will take any position on its own curvature.

Consider the cutting of an arch with a 300 mm face. In addition to the voussoir templet, two others are required (Fig. 9.56). These are called the 200 mm and 100 mm templets. The arch must be cut to a definite system. For instance:

1. Cut the soffit bevels on all the bricks, stacking as shown in Fig. 9.57.
2. On the back face of each brick, scribe its particular number so that it will be placed in its correct position.
3. Reduce the bricks to their voussoir size.

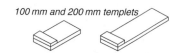

Figure 9.56 Header and stretcher templets for camber arch

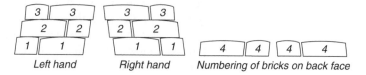

Figure 9.57 Camber arch voussoirs stacked by numbers

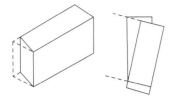

Figure 9.58 Marking stretcher voussoirs

In operation 1 above, apply the bevel which has previously been set to its number, scribe with a pencil to give a clear cutting line, and cut. To obtain the header face, apply the 100 mm templet, and for the stretcher face apply the 200 mm templet (Fig. 9.58).

In operation, when ready to reduce the bricks to voussoir size, fix the templet to a rigid flat base. It will be found convenient to make the fixings coincide with the cutting marks (Fig. 9.59). Place the bricks to the templet, scribe, and cut. Stack neatly in the manner illustrated (Fig. 9.60).

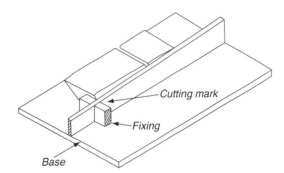

Figure 9.59 Reducing or rubbing bricks to voussoir size

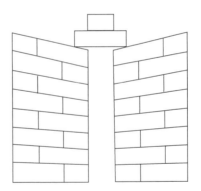

Figure 9.60 Stacking finished voussoirs

Setting or fixing of arches

Types of support

No attempt will be made under this heading to show the type of temporary arch supports over large spans. Only those met with by the bricklayer in the course of everyday work will be dealt with.

Turning piece is made from solid timber for arches of limited span. For an arch with a 225 mm soffit, two 75 mm timbers can be placed side by side (Fig. 9.61). Alternatively it could be constructed from plywood and packed out with timbers (Fig. 9.62).

Open-lagged centre is framed from light timber and used for the turning of ringed arches (Fig. 9.63).

Close-lagged centre is suitable for the turning of an axed or gauged arch. Its use facilitates the marking of the voussoir positions (Fig. 9.64).

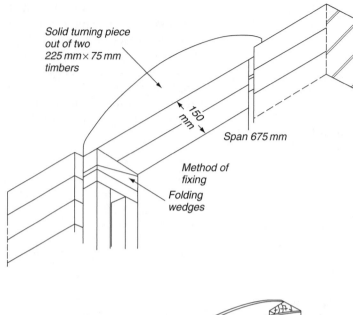

Solid turning piece
out of two
225 mm × 75 mm
timbers

150 mm

Span 675 mm

Method of
fixing

Folding
wedges

Figure 9.61 Setting up arch
turning pieces

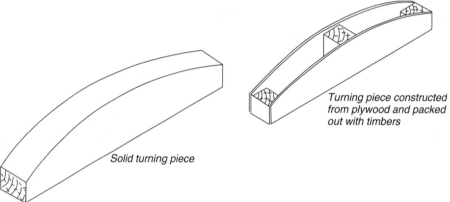

Turning piece constructed
from plywood and packed
out with timbers

Solid turning piece

Figure 9.62 Solid turning piece and built up type

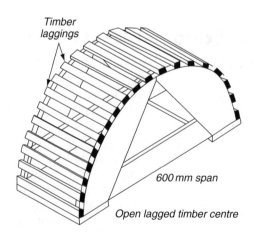

Timber
laggings

600 mm span

Open lagged timber centre

Figure 9.63 Open lagged
timber centre for semi-circular
arch

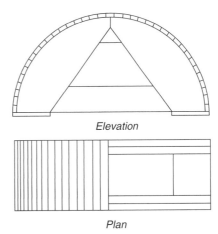

Elevation

Plan

Figure 9.64 Closed lagged timber centre for semi-circular arch

When using temporary arch supports, folding wedges are placed directly under the centre. This facilitates the removal of the centre or turning piece and, in particular, avoids the chipping of the arch face on the intrados arris, which invariably occurs if wedges are not used.

Modern methods

There are also new methods of supporting arches in both solid and cavity walls. These arch formers are available in plastic and steel. The centres are built in and remain as part of the structure.

A semi-circular centre is shown but segmental centres are also available (Fig. 9.65).

Another version which sits on a steel lintel is shown in Fig. 9.66.

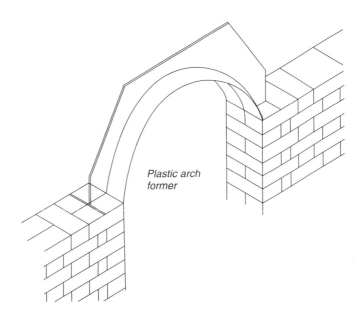

Plastic arch former

Figure 9.65 Plastic arch former for solid walls

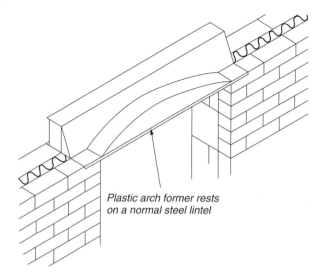

Plastic arch former rests
on a normal steel lintel

Figure 9.66 Plastic arch
former sitting on a boxed steel
lintel on cavity walls

Preparation for setting

Consider the setting of a semi-circular arch:

Fix and adjust the temporary arch support. At the striking point on the centre fix a short length of bricklayer's line for checking that every voussoir is radiating correctly to the striking point when it is bedded (Fig. 9.67). With a pair of dividers, mark on the arch centre the correct positions of all the voussoirs.

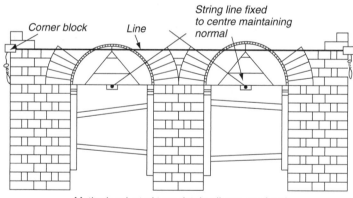

Corner block   Line

String line fixed
to centre maintaining
normal

Figure 9.67 Setting arches

Methods adopted to maintain alignment of arch

To keep the arch in alignment, various methods can be adopted, e.g. if the arches are on a small flank wall, as in Fig. 9.67, the corners can be erected, using line and pins. If, on the other hand, the arch is one of many occurring on a large frontage, then a good method is to use profiles as Fig. 9.68.

For the cutting of skewbacks in preparation for the fixing of axed segmental or camber arches, always use a 'gun' or templet. Its application is shown in Fig. 9.69, and its function is to keep the angle of the skewback constant. It often happens that several bricklayers are engaged in building a long frontage and each is required to cut the skewbacks in a particular stretch

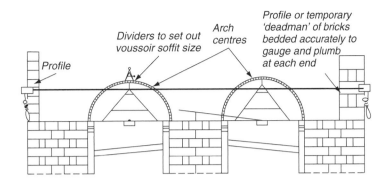

Figure 9.68 Alternative method to maintain alignment of arches

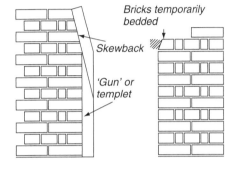

Figure 9.69 Use of gun or templet to obtain correct skewback angles

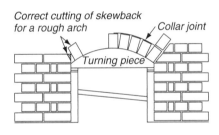

Figure 9.70 Alternative method of marking skewback angle for rough arch

of walling. If each is given the set-back of the skewback and then left to work independently, errors will occur, but if provided with 'guns' previously prepared by the foreman bricklayer and used in the proportion of, say, one gun to each group of three or four bricklayers, errors will be avoided.

For the cutting of a skewback for a ringed arch no gun is needed, as the turning piece can be fixed and the skewback ascertained from this by placing a brick flat on the turning piece, as shown in Fig. 9.70.

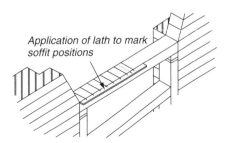

Figure 9.71 Marking voussoirs on camber arch support timber

Setting a camber arch

Having cut the skewbacks and fixed the turning piece, preparation is made for the marking of the voussoir positions. To do this, take a lath and place it on the drawing or setting out of the arch, transfer the soffit marks to the lath, and apply to the turning piece, working right and left from the centre (Fig. 9.71). Note the reason for additional traversing to obtain

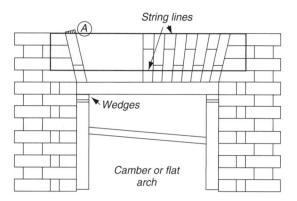

Figure 9.72 Setting camber arch voussoirs with string lines top and bottom

precise voussoir sizes, as previously described. Use the line and pins (Fig. 9.72) to maintain alignment. In cutting a camber arch, the top portions of bricks occurring on the extrados are often uncut and adjusted after being fixed or set, see point A in Fig. 9.72. Note that the soffit of the

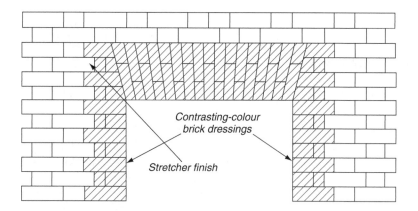

Figure 9.73 Overall rectangular finish of dressing must be anticipated when setting out facework at ground level

arch is cambered but that the extrados is horizontal so as to conform with the general brickwork.

Special attention should be given to the bond setting out of general brickwork where a different coloured dressing occurs on the reveals and a camber arch is used to cover an opening. Most architects prefer the dressings to finish at the top of the arch, as shown in Fig. 9.73. The appearance is rectangular and not as illustrated in Fig. 9.74. The arrangement of bricks at the setting out level, which will probably be at ground level, must be watched closely so that proper sequence is obtained.

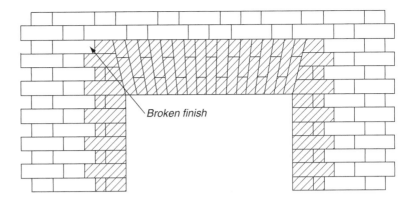

Figure 9.74 Alternative appearance to Fig. 9.73

Circular Bull's-eye

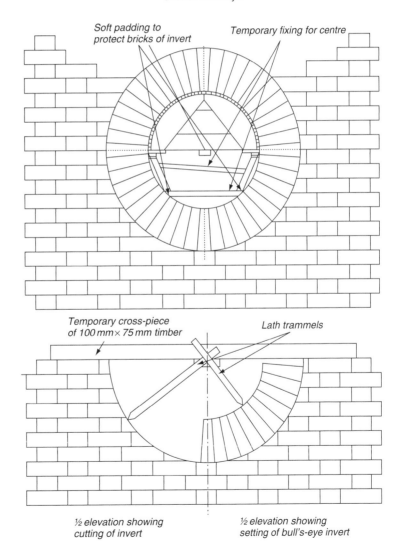

Figure 9.75 Circular bull's-eye

Bull's eye (Figs 9.75 to 9.77)

Although the detailed study of the construction of a bull's-eye belongs to the more advanced stage of brickwork, its construction will be dealt with in this chapter.

The setting out and cutting of an axed bull's-eye and the setting or fixing of this and the ringed bull's-eye differ slightly from the ordinary arch in that the lower half or invert requires no temporary supports, but needs the aid of a trammel for its formation in the wall (lower half of Fig. 9.75). Note that the key brick of the invert is laid first and that the voussoir positions must

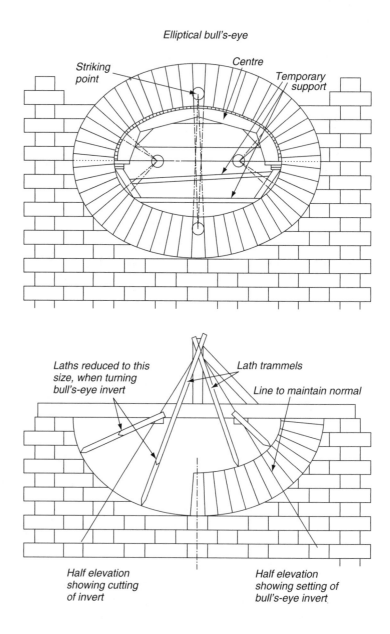

Figure 9.76 Elliptical bull's-eye with major axis horizontal

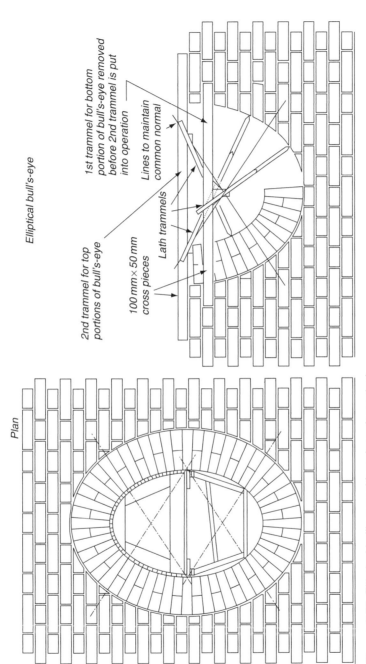

Elliptical bull's-eye

1st trammel for bottom
portion of bull's-eye removed
before 2nd trammel is put
into operation

Lines to maintain
common normal

2nd trammel for top
portions of bull's-eye

100 mm × 50 mm
cross pieces

Lath trammels

Plan

Figure 9.77 Elliptical bull's-eye with major axis vertical

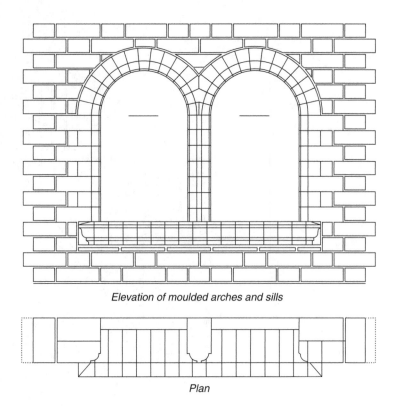

Elevation of moulded arches and sills

Plan

Figure 9.78 Intersecting moulded arches

be marked off from its extrados, and not marked off from its intrados, on the centre or turning piece, as in the case of the arch.

The illustrations which follow will introduce some craft terms. From the information already given, the apprentice will be able to follow the craft operations involved in their erection.

Intersecting semi-circular arch (Fig. 9.78)

Gauged half-brick intersecting semi-circular arch, with moulded intrados to conform with moulded reveals. There is a half-brick moulded sill made up to courses by the introduction of a tile course.

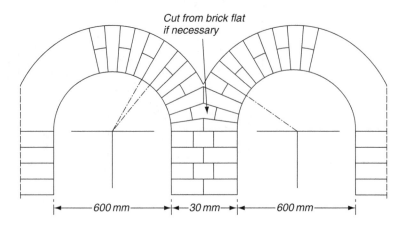

Cut from brick flat
if necessary

|←——600 mm——→|←—30 mm—→|←——600 mm——→|

Figure 9.79 Intersecting axed arches, 225 mm bonded face

Figure 9.79 shows the intersection of two semi-circular arches, 225 mm on face, on a 330 mm pier. Note the cutting of the brick in the second course from springing.

Arches with basketweave brick core or tympanum

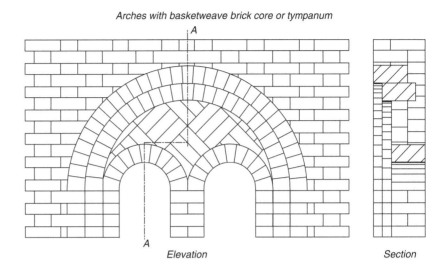

Figure 9.80

Elevation   *Section*

Arch with a brick core or tympanum

The brick core can be erected in two ways:

1. By the use of a trammel, using the core as a support to the main arch during construction.
2. By first erecting the main arch on a temporary support or centre and inserting the smaller arches and brick filling when the temporary support is removed (Fig. 9.80). Two examples are shown in Figs 9.80 and 9.81.

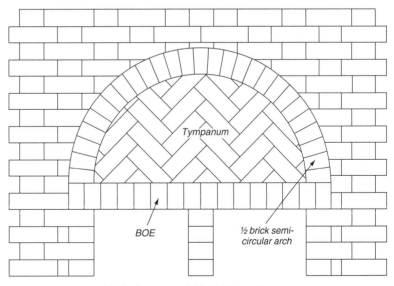

Figure 9.81 Introducing ornamental brick filling or tympanum

Introducing ornamental brick filling or tympanum

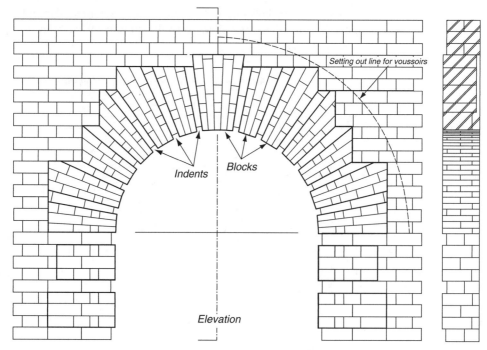

Figure 9.82 Indented arch

Indented arch (Fig. 9.82)

Sometimes termed a 'blocked' or 'rusticated' arch. The block on reveals and arch are usually in a different coloured brick, an attractive combination being that of a sand-faced multi-red brick for the blocks with a silver-grey facing brick for the general brickwork. The courses between each block are in the same colour as the general brickwork.

Note the method of obtaining correct sizes of voussoirs, by constructing a curve concentric to the intrados from the greatest depth of arch face.

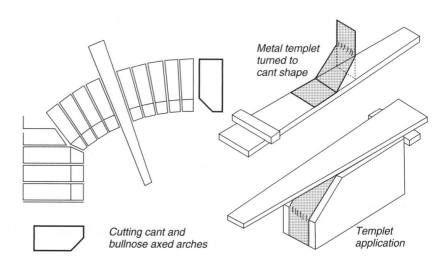

Figure 9.83

Cant brick arch (Fig. 9.83)

Shows the method of cutting axed arches constructed of cant or bullnose bricks.

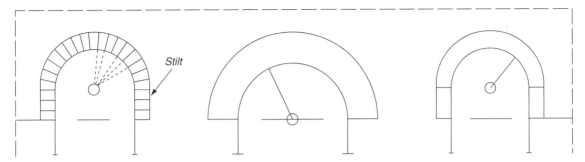

Figure 9.84 Use of stilted arches to make crowns level with larger span arch

Stilted arches (Fig. 9.84)

Although this type of arch was used in isolated cases for decorative purposes, it can be and was originally used to maintain alignment at the crowns of arches where small openings occurred together with large openings.

Horse-shoe arch (Fig. 9.85)

The origin of this arch was probably oriental and the crown of the arch was often pointed instead of circular, and in such a case the arch was known as 'Moorish'.

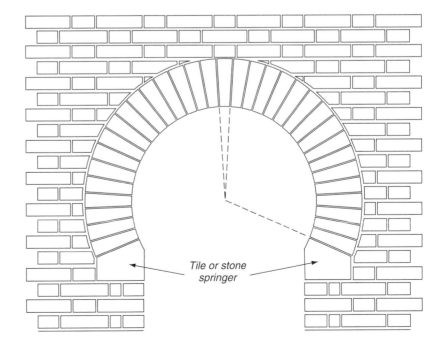

Figure 9.85 Horse-shoe arch

Arches in orders

In these the brickwork was recessed in 'orders', as shown in Fig. 9.86, and in the early days they occurred in very thick walls. Today, orders do not necessarily occur in thick walls, and can be created for decorative purposes in external cavity wall construction. A probable reason for their use in

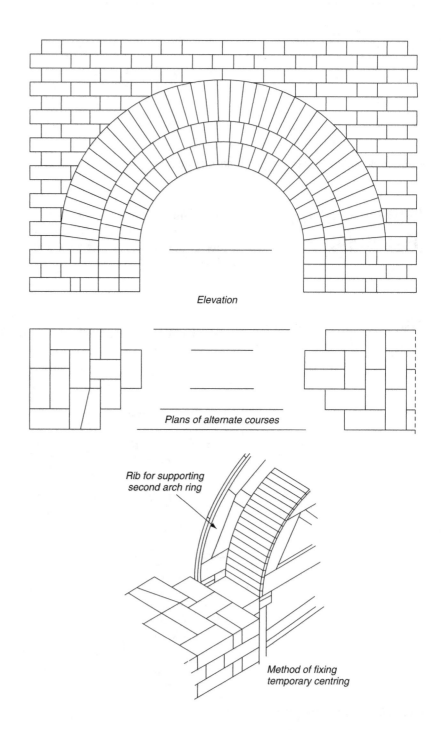

Elevation

Plans of alternate courses

Rib for supporting second arch ring

Method of fixing temporary centring

Figure 9.86 Construction of recessed arches

early building was the difficulty of obtaining and constructing temporary supports: by recessing, only a temporary support for the inner arch ring was necessary, successive orders being corbelled. In modern practice, where the orders project 112.5 mm, it is advisable to use temporary supports for every projection.

This type of arch was originally adopted to overcome difficulties in construction. Other forms of opening also occurred in thick external walls; the width on the external face might be narrow while in the internal face it was broad, often with splayed jambs. Several reasons are given for their use; for instance, a narrow arch on the external face of the wall would prevent the entry of thieves, glass being either unknown or not available, while a wide arch on the inner face would give light; again such orders were constructed in castles to protect the archers withstanding a siege.

Arches in orders on splayed jambs

Figure 9.87 illustrates this type of arch. Care must be taken when erecting the temporary ribs used for the turning of the separate rings to see that

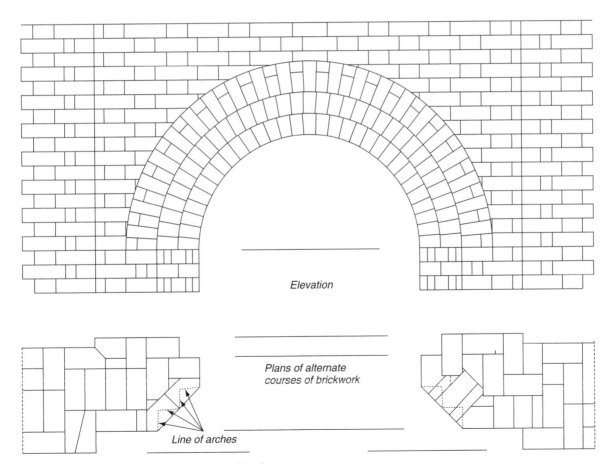

Elevation

Plans of alternate
courses of brickwork

Line of arches

Figure 9.87 Recessed arches on splayed jambs

these can be easily removed after construction. If notches are not cut at the springing of the ribs where they rest on the recess, it is obvious that they will bind and will be difficult to remove. This will tend to spall the edges of the arch soffit and also the top of the brick jamb. The recess occurring between the arch and jamb should be weathered by some suitable method, e.g. by tile creasing built in or by continuance of the jambs until they intersect the soffit of one ring of the arch and the face of the other.

Rear arch

Figures 9.88 to 9.91 show the plan, elevation, and section of a reararch. This is a segmental pointed arch and it will be noted that the jambs of the opening are splayed but the soffit of the arch is level. It is considered that the best method of construction is shown in Fig. 9.88 (Method 1). The skewback is cut through the abutment and the splayed jamb is 'made good' to the soffit of the arch after it has been built and the temporary centring removed.

Figure 9.89 (Method 2) might be adopted, and in this case the intersection between the splayed jambs and the level soffit of the arch must be developed in order to gain the true shape of the templet for cutting the skewback. Figure 9.90 illustrates the method of obtaining the true shape of intersection. The length of the splayed jamb in plan is divided in this case into eight equal parts and the points projected vertically into the elevation. The distances from the line xy, i.e. $a0$ . . . $b1$ . . . $c2$ . . . $d3$ . . . etc., are taken from the elevation (Fig. 9.91).

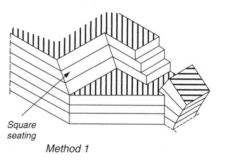

Square seating

Method 1

Figure 9.88 Cutting a normal skewback first, then making good splayed jambs after arch centre is removed

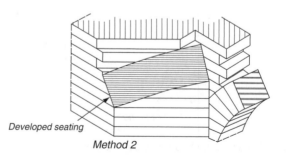

Developed seating

Method 2

Figure 9.89 Formation of a mitred skewback between arch and bearing

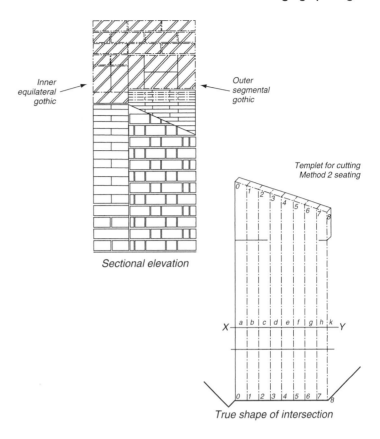

Inner equilateral gothic

Outer segmental gothic

Templet for cutting Method 2 seating

Sectional elevation

X — a b c d e f g h k — Y

True shape of intersection

Figure 9.90 Reveal of rear arch

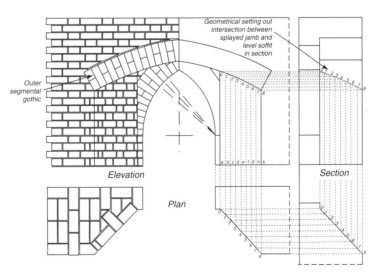

Geometrical setting out intersection between splayed jamb and level soffit in section

Outer segmental gothic

Elevation

Plan

Section

Figure 9.91 Rear arch

Bonnet arch

Figure 9.92 shows an axed arch where tiles have been introduced on the inner arch because of the thin nature of voussoirs that would have occurred at this point. A sound method to adopt in cutting this arch is to

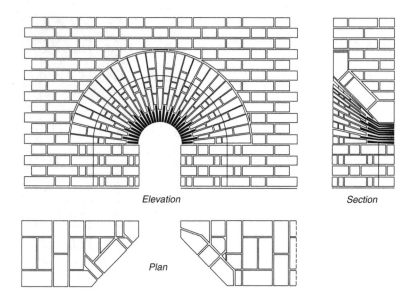

Elevation   *Section*

Plan

Figure 9.92 Bonnet arch

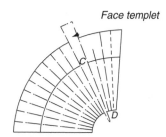

Figure 9.93 Elevation of bonnet arch showing tapering voussoirs on face

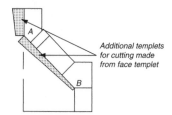

Figure 9.94

Figure 9.95 Figure 9.94 templets fixed in preparation for cutting voussoirs

mount the templet for reducing bricks to voussoir shape on a flat board. Procedure for setting out and cutting:

1. Set out the face templet in the usual way (Fig. 9.93).
2. Prepare the templet for mounting, noting that the templet for the splay is elongated and that the points A and B in Fig. 9.94 must equal points C and D in Fig. 9.93.
3. Mount the templet and prepare bricks for reducing Fig. 9.95. The squint brick must be bevelled to the arch curvature before being reduced; other bricks can remain square. Having placed the bricks to the mounted templet, scribe and cut.

It will be realised that the arch, besides being wedge-shaped on the face and splayed soffit, is also wedge-shaped in its thickness, and the method of cutting as explained neglects allowance for this. In spite of this, the method described is simple and avoids complications, the size of the wedge in thickness of the arch is very small and the difference can be made up in mortar on the back edge.

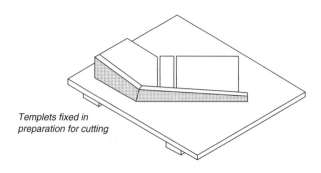

Templets fixed in preparation for cutting

10 Cavity walling and brick cladding

Definition

The standard form of construction for the external walls of brick buildings is called cavity walling. This means that the bricklayer builds the two separate 'leaves' or 'skins' of 'brick masonry' (a general term indicating brickwork and/or blockwork) with a 50 mm to 75 mm wide space between. The outer skin is usually 102.5 mm thick face brickwork, but may be constructed from facing quality blocks. The inner skin is usually 100 mm thick common blocks that are later plastered to receive internal decoration (see Fig. 10.1).

Both skins of brick masonry are joined together with a regular pattern of corrosion resistant ties, so that they behave as one single wall.

Reason for the title

The reasons for the double title of this chapter are that from the bricklayer's point of view, the craft skill requirements for cavity wall brickwork and brick cladding are very similar.

First, cavity wall brick masonry may be load bearing, that is, used to support the load from floors and roof. This application is usually confined to low rise buildings of up to three to four storeys only (see Fig. 10.2).

Second, identical looking cavity walling is widely used to cover the framed structure of high rise/multi-storey building, where it is not required to support the load of floors and roof, and is referred to as brick cladding (see Fig. 10.3).

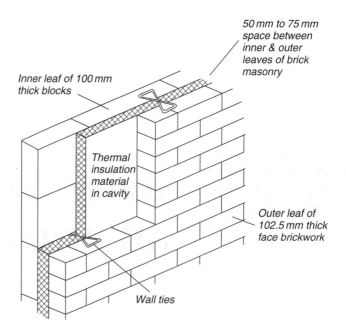

50 mm to 75 mm space between inner & outer leaves of brick masonry

Inner leaf of 100 mm thick blocks

Thermal insulation material in cavity

Outer leaf of 102.5 mm thick face brickwork

Wall ties

Figure 10.1 Typical insulated cavity walling

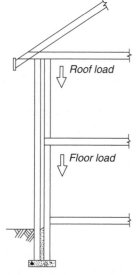

Roof load

Floor load

Figure 10.2 Load bearing cavity walling

Third, a 102.5 mm thick wall (single skin) of face brickwork may be used to cover large pre-cast concrete panels of some types of industrial buildings or in-situ concrete walling and bridge parapets (Fig. 10.4). Special fixings are required to fasten the cavity brickwork to the pre-cast concrete.

Fourth, a single skin of 102.5 mm face brickwork is used as brick cladding to timber frame houses. Here the inner leaf is formed from factory-made storey height panels of structural timber which support floor and roof loads (Fig. 10.5). Special cramps are used to tie the brickwork to the timber frame.

All the craft skill requirements of the bricklayer contained in this chapter refer equally to these four different applications of cavity walling.

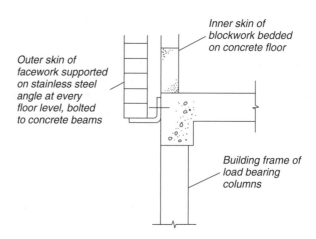

Inner skin of blockwork bedded on concrete floor

Outer skin of facework supported on stainless steel angle at every floor level, bolted to concrete beams

Building frame of load bearing columns

Figure 10.3 Non-load bearing cavity wall cladding to framed structure

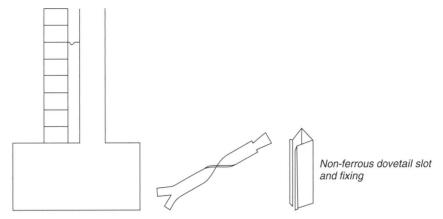

Figure 10.4 Non-load bearing cavity wall cladding to concrete structure. Special dovetail slots and fixings are required

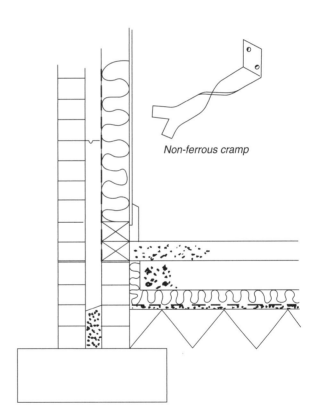

Figure 10.5 Non-load bearing cavity wall cladding to timber framed structure

Purpose

Cavity wall construction began to be widely used from the 1920s as a way of preventing dampness from soaking through the outer walls of buildings. The 50 mm to 75 mm wide gap stopped rain penetrating from the outer surface to the plastered inner surface of external walls by capillarity (when water is drawn through hairlike channels within a porous structure, a brick say, by the action of surface tension). This was possible when outer walls were commonly 215 mm thick solid brickwork.

As a secondary advantage, this space between the inner and outer skins also provides thermal insulation for modern buildings, because heat energy mainly escapes by conduction through solid material. Air is a poor conductor of heat energy, therefore the rate of heat loss is very much slower than was the case when buildings had solid outer walls (see Fig. 10.6).

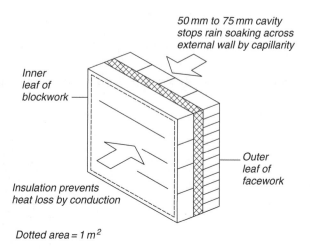

Figure 10.6 Function of typical cavity walling

Building Regulations

The basic advantages of cavity wall construction, over solid 215 mm thick brickwork, for the outer walls of a building have been incorporated in the current Building Regulations. In order to satisfy requirements of the current Building Regulations and enable planning permission to be obtained, cavity wall construction is usually specified whether the structure is low rise or multi-storey.

Figure 10.7 indicates how basic requirements of the Building Regulations are satisfied, where standard strip foundations are specified with a solid ground floor slab.

Figure 10.8 shows a trench-fill foundation, associated with a suspended ground floor construction of pre-stressed, pre-cast concrete floor beams supporting standard size concrete blocks.

A External cavity wall, providing resistance to through-penetration of rain
B Horizontal dpc in both leaves, not less than 150 mm above ground level, to prevent dampness rising from the soil
C A minimum distance of one metre between ground level and the underside of the concrete foundation, as a protection against frost heave in winter and drying shrinkage of clay subsoils in summer
D A minimum 150 mm thickness of foundation concrete to transfer the building load adequately on to the 'natural foundation' of the subsoil
E Sulphate resisting cement in the foundation concrete and substructure brickwork, necessary where soluble sulphates are present in subsoil water

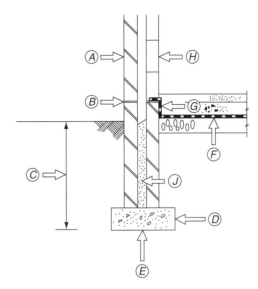

Figure 10.7 Cavity walling on standard strip foundation indicating current Building Regulation requirements with solid ground floor

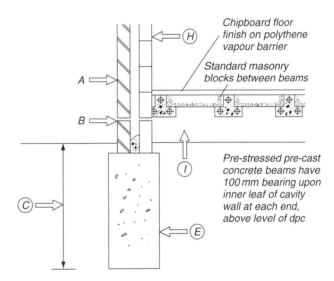

Figure 10.8 Cavity walling on trench fill foundation with suspended concrete floor

F   Continuous damp proof membrane (dpm) across the ground floor areas to prevent dampness rising from the soil

G   Dpm and dpc, lapped and joined within the floor thickness around the perimeter of rooms

H   External walls built of lightweight blocks and other thermal-insulating material to give a 'U' value of 0.45. (In other words, heat energy must not escape through the outer walls at a rate greater than 0.45 watts per square metre, per hour, per degree difference in temperature internally and externally)

I   A ventilated air space separates the suspended floor from damp soil

J   Weak cavity fill to prevent collapse of walls due to pressure from the ground

Function

Figure 10.2 shows that it is the inner leaf of cavity walling that largely supports the load from floors and roof in a low rise building of load bearing wall construction. Common building blocks have totally replaced common bricks for this inner leaf, due to improved thermal insulation values and also bricklayer output (i.e. one standard size 100 mm thick block is equal to six bricks).

The purpose of the outer skin of facework is to give the building a weather resistant and pleasant appearance, by selection from the wide range of colours and surface textures of bricks and facing blocks available to the designer.

Although cavity walling is constructed with properly finished, solidly filled mortar joints, it is expected that this outer skin will let rainwater soak through as far as the cavity. This is because mortar, bricks and blocks are porous to varying degrees. This through-penetration will be highest on those elevations of a building exposed to prevailing wet winds.

Where cavity walling is used as cladding to a high rise building the degree of exposure to wind driven rain increases with height.

Thermal insulation

When cavity walling was first used, it was common practice to install air bricks at the top and bottom, at intervals around the whole perimeter of the building, to ventilate this 50 mm to 75 mm wide space to remove damp air. Since the 1950s, however, cavity walls have become sealed, with insulating material built in as work proceeds, to improve the thermal insulation value of the cavity space.

Two ways that a bricklayer may be told to install thermal insulation in cavity walling are shown in Figs 10.9 and 10.10. The fully filled system is where flexible fibre 'batts' of insulation completely occupy the cavity space.

The partial fill system is where stiffer 'boards' of insulating material half-fill the cavity, but retain a 25 mm to 35 mm air space as well. Both batts and boards are supplied in purpose-made sizes to fit neatly between layers of wall ties and are approximately 900 mm in length.

All insulation batts should be fixed staggering the vertical joints, and butting the vertical and horizontal joints as tightly as possible.

Insulation batts can be cut with a sharp knife but always ensure the cut is square and forms a perfect tight joint.

Insulation batts for full fill insulation are available in various thicknesses and can be made from mineral fibre which is soft and flexible.

Insulation for partial cavity fill is made from more rigid insulation boards such as expanded polystyrene bead board.

The boards in partial cavity fill should only be fixed with special wall ties with plastic clips to hold the insulation back against the internal blockwork.

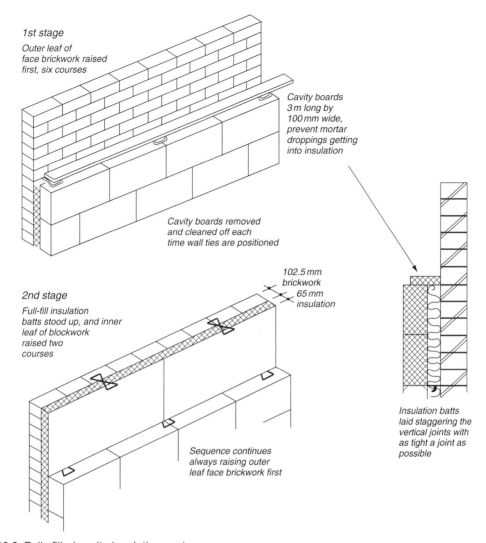

1st stage

Outer leaf of
face brickwork raised
first, six courses

Cavity boards
3 m long by
100 mm wide,
prevent mortar
droppings getting
into insulation

Cavity boards removed
and cleaned off each
time wall ties are positioned

102.5 mm
brickwork
65 mm
insulation

2nd stage

Full-fill insulation
batts stood up, and inner
leaf of blockwork
raised two
courses

Sequence continues
always raising outer
leaf face brickwork first

Insulation batts
laid staggering the
vertical joints with
as tight a joint as
possible

Figure 10.9 Fully filled cavity insulation system

Bonding

Cavity wall construction has been responsible for the disappearance from modern buildings of the many and various traditional bonding patterns which are possible with the use of headers and stretchers. This is because use of the full range of face bonds requires walls to be at least 215 mm thick if headers are to be used effectively.

Stretcher bond is best suited to the 102.5 mm thick outer skin of cavity walling and for making the most economic use of the longer stretcher face of each expensive facing brick; see Fig. 10.11. However if another face bond is required for cavity walling, to match existing work, then it is possible to use 'snapped headers' (half bats) as shown in Fig. 10.12. This application could also produce English bond.

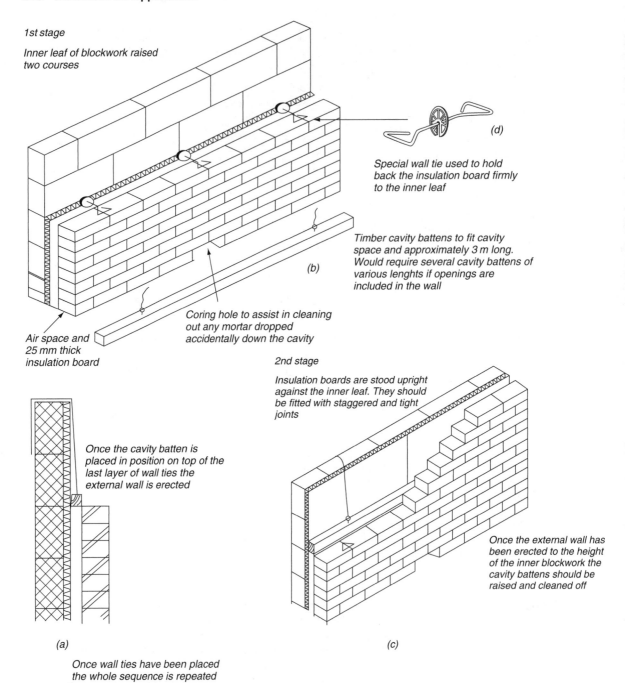

1st stage

*Inner leaf of blockwork raised
two courses*

(d)

*Special wall tie used to hold
back the insulation board firmly
to the inner leaf*

*Timber cavity battens to fit cavity
space and approximately 3 m long.
Would require several cavity battens of
various lenghts if openings are
included in the wall*

(b)

*Coring hole to assist in cleaning
out any mortar dropped
accidentally down the cavity*

*Air space and
25 mm thick
insulation board*

2nd stage

*Insulation boards are stood upright
against the inner leaf. They should
be fitted with staggered and tight
joints*

*Once the cavity batten is
placed in position on top of the
last layer of wall ties the
external wall is erected*

*Once the external wall has
been erected to the height
of the inner blockwork the
cavity battens should be
raised and cleaned off*

(a)

(c)

*Once wall ties have been placed
the whole sequence is repeated*

Figure 10.10 Partially filled cavity insulation system

Wall ties

Headers and courses of headers in solid brick walling are there to tie the wall from back to front (see Chapter 4, Bonding Rule 2). In cavity wall constructions headers cannot be used for this purpose, because they would allow dampness to cross the cavity by capillary action.

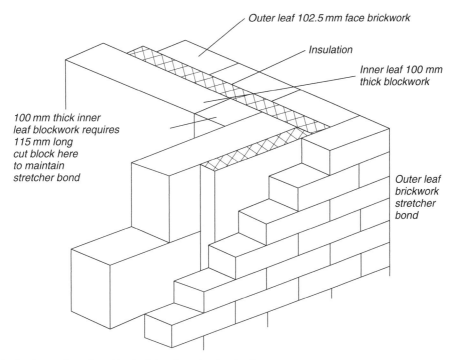

Outer leaf 102.5 mm face brickwork

Insulation

Inner leaf 100 mm thick blockwork

100 mm thick inner leaf blockwork requires 115 mm long cut block here to maintain stretcher bond

Outer leaf brickwork stretcher bond

Figure 10.11 Cavity wall quoin with stretcher bond outer leaf

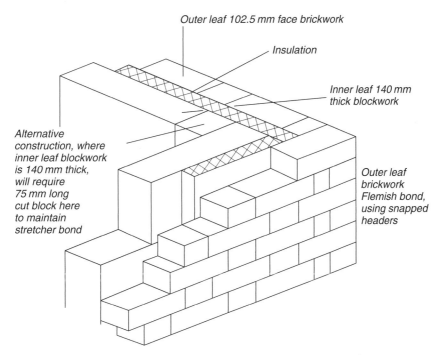

Outer leaf 102.5 mm face brickwork

Insulation

Inner leaf 140 mm thick blockwork

Alternative construction, where inner leaf blockwork is 140 mm thick, will require 75 mm long cut block here to maintain stretcher bond

Outer leaf brickwork Flemish bond, using snapped headers

Figure 10.12 Cavity wall quoin, showing use of snap headers in 102 mm outer leaf

*Double triangle
with plastic clips*

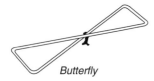

Plastic

Double triangle

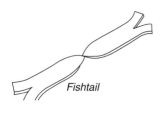

Fishtail

Butterfly

Stainless steel

Figure 10.13 Different types of wall ties

Figure 10.14 Basic requirements for cavity wall ties

Therefore, a range of proprietary ties are made for this job of tying together inner and outer skins of brick masonry, as 'substitute headers', so that both leaves behave as one wall. These wall ties are made from stainless or galvanised steel, or polypropylene so they do not provide a passage for moisture.

The general and important requirements for any pattern of wall tie are as follows (see Fig. 10.14):

A It must be made durable and non-corroding.
B It must have a central drip or twist to prevent water from tracking across the cavity.
C When laying it, allow for not less than 50 mm embedment in each leaf of the cavity walling.
D It must have a particular end shape, to give a secure grip in the bed-joint mortar.
E A tie of the correct overall length must be used, to span the cavity width plus two embedments.

The standard maximum spacing for wall ties is at intervals of 900 mm horizontally and every sixth course vertically. Each horizontal layer should be offset (see Fig. 10.15). For purposes of estimating quantities of wall ties required, this works out at approximately 2.5 per m².

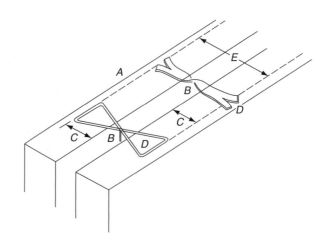

Cleanliness

For cavity walling to be effective, wall ties, insulation and cavity gutters must be kept free of mortar droppings as work proceeds. If, due to carelessness and poor supervision, cavities are not kept clean, then dampness will be able to cross the cavity through porous mortar droppings.

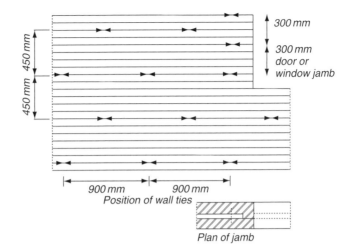

Figure 10.15 Standard spacing of wall ties (maximum distances apart)

Cavity battens or boards, approximately 3 m long, raised and cleaned off every six courses as work proceeds, are the best way of preventing mortar droppings falling into cavity walling (see Fig. 10.16).

Alternatively, where fully filled cavity insulation is specified, plain battens without lifting wires are used; see Fig. 10.17.

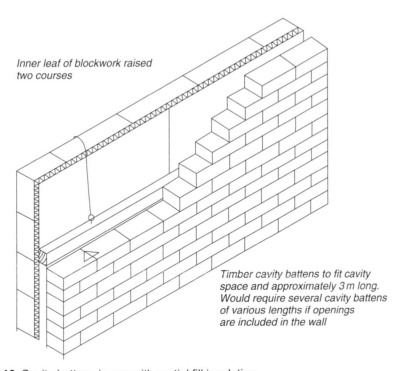

Figure 10.16 Cavity battens in use with partial fill insulation

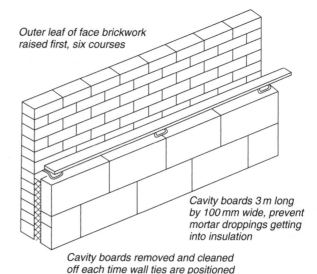

Outer leaf of face brickwork
raised first, six courses

Cavity boards 3 m long
by 100 mm wide, prevent
mortar droppings getting
into insulation

Figure 10.17 Cavity battens in
use with full filled insulation

Cavity boards removed and cleaned
off each time wall ties are positioned

Coring holes

It is usual to leave temporary openings, called coring holes, in the outer leaf, over all cavity trays. These holes, of one-brick size, are for the removal of any mortar droppings, jointers, etc., that have got past the cavity battens; see Fig. 10.18(a). Cavity ties and trays should be inspected

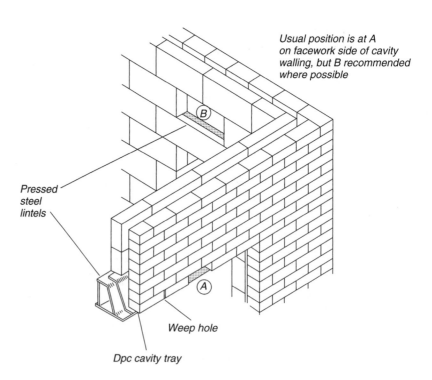

Usual position is at A
on facework side of cavity
walling, but B recommended
where possible

B

Pressed
steel
lintels

A

Weep hole

Figure 10.18 Location
of coring holes

Dpc cavity tray

and cleaned in this way at the end of every day's work. Where possible, however, it is better if the one-block sized coring holes be left out of the inner leaf; see Fig. 10.18(b). This is recommended, as it provides a bigger temporary opening. It also avoids the risk of the slight difference in mortar colour which would highlight the coring holes in the finished facework. The coring holes are sealed up when the external scaffolding is removed.

Temporarily, bedding bricks or blocks in sand at points A or B provide support for those above, and make for easy removal to form the coring holes.

Coring holes should only be regarded as a back-up procedure, and not a substitute for cavity battens or boards. See also Fig. 10.10(a) showing coring hole at ground level.

Openings

Leaving a door or window opening in a solid brick wall is a relatively simple job, as it merely requires making allowance for the correct space for a frame, plus a lintel or arch across at the right height.

Leaving the same opening in a cavity wall, however, is a much more complicated process. Detailed provision must be made for dealing with:

1. dampness, which may soak through the sides or reveals of any opening (see Fig. 10.19);
2. rainwater, which may run down inside the cavity from above the opening (see Fig. 10.24);
3. rain, which may soak through any sill or threshold of an opening (see Figs 10.20 to 10.24).

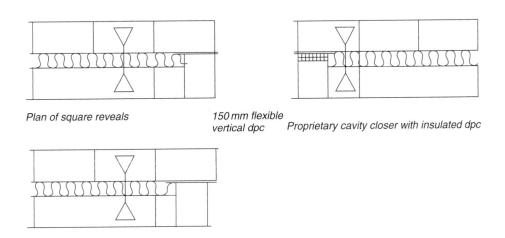

Plan of square reveals

150 mm flexible vertical dpc   Proprietary cavity closer with insulated dpc

Plan of recessed reveals   (a)

Figure 10.19 Location of vertical dpc at reveals

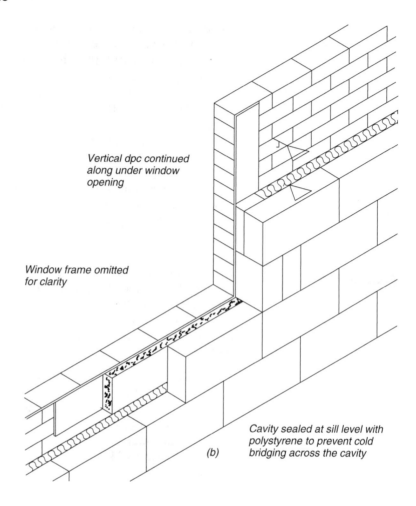

Vertical dpc continued along under window opening

Window frame omitted for clarity

Cavity sealed at sill level with polystyrene to prevent cold bridging across the cavity

(b)

Figure 10.19 *(Continued)*

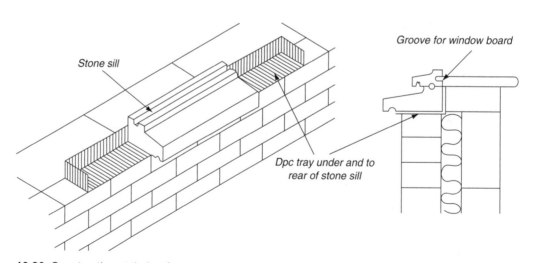

Stone sill

Groove for window board

Dpc tray under and to rear of stone sill

Figure 10.20 Construction at timber frame on stone sill showing cavity tray

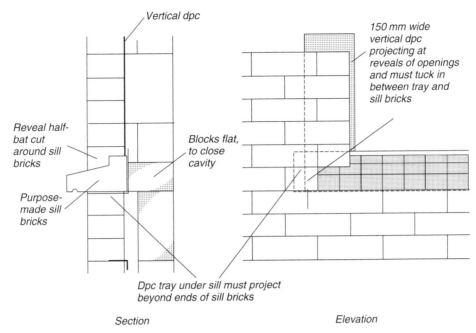

Figure 10.21 Purpose-made sill bricks at window opening

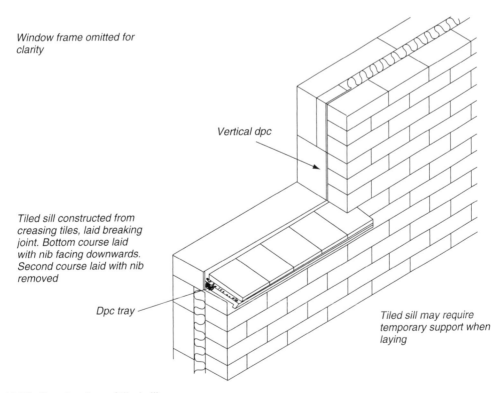

Figure 10.22 Construction of tiled sill

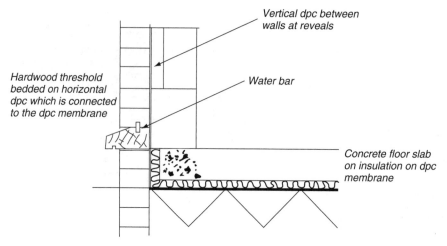

Figure 10.23 Damp proofing treatment at door threshold

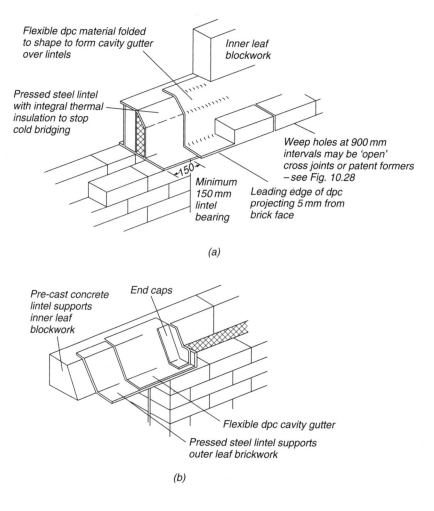

Figure 10.24 (a) Typical cavity tray or gutter. (b) Purpose-made end caps glued to cavity tray

DPC cavity trays

Bearing in mind that rainwater will penetrate the outer skin of brickwork, trays or gutters of flexible bituminous felt must be built into cavity walls to collect the water and prevent dampness from reaching the inner skin of blockwork; see Fig. 10.24(a), showing patent steel lintel bridging an opening.

These cavity gutters of dpc material are needed above every opening in cavity wall construction and at every floor level of a multi-storey building, where cavity walling is used as the external cladding.

Purpose-made stop ends of dpc material (see Fig. 10.24(b)) are glued at either end, to prevent run-off into wall insulation and to encourage draining through weep holes.

'Running laps' at the end of one roll of dpc and the next, including internal and external angles, must be effectively sealed and jointed, so as to remain watertight.

Purpose-made polyethylene units may be specified to provide permanent support to running laps in dpc cavity gutters, as shown in Fig. 10.27.

Rolls of dpc should always be stored on end to avoid squashing and distortion, which will make the material more difficult to flatten out when used. In cold weather, the rolls should be stored in a warmer place to make bedding the material that much easier.

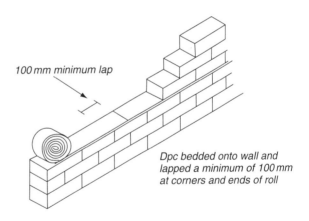

100 mm minimum lap

Dpc bedded onto wall and lapped a minimum of 100 mm at corners and ends of roll

Figure 10.25 Lap jointing flexible dpcs

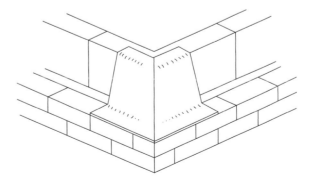

Figure 10.26 Purpose-made dpc tray corner unit

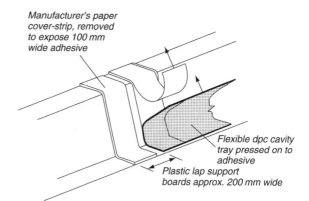

Figure 10.27 Patent rigid plastic support board for running laps in dpc trays

Weep holes

Empty or 'open' cross joints are left at intervals of 900 mm in the outer leaf of cavity walling, at the level of cavity gutters, to act as permanent drainage points. These 'weep holes' may simply be left as a cross joint free of mortar, or formed with plastic or nylon fibre inserts.

Both types of plastic or nylon fibre inserts are available in a limited number of colours to match mortar joint colour. Where a light colour mortar is specified, open weep holes appear black and will be conspicuous if not in matching locations above and below every window opening. With the external elevations of a building, it is important that the bricklayer takes care to keep the pattern of weep holes under and over openings symmetrical; see Fig. 10.28.

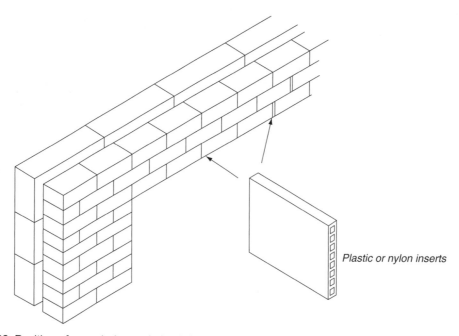

Figure 10.28 Position of weep holes and plastic inserts

Cavity details at eaves level

To prevent birds that get into the roof spaces from nesting in wall insulation, it is usual to close or seal off cavity walls at the top. Blocks laid flat is a convenient way of doing this. This also spreads the load of a pitched roof between both inner and outer skins of brick masonry; see Fig. 10.29. The roof rests on timber wall plate which has galvanised restraint straps to secure it to the inner leaf.

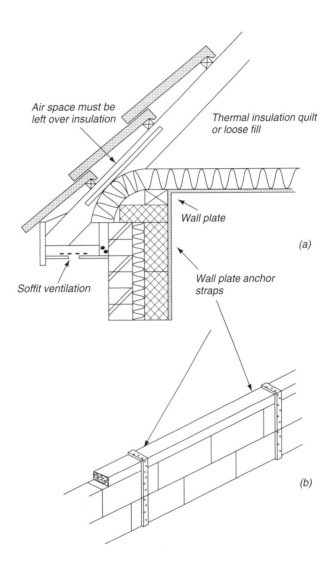

Figure 10.29 Sealing cavities at eaves level

Cavity walls below ground level

As an aid to stabilising a cavity wall below ground level the cavity is filled with weak concrete up to ground level. The top of the concrete filling is sloped towards the outer face wall to ensure any water dripping down the cavity escapes through the weep holes. The insulated cavity should extend at least 150 mm below the lowest horizontal dpc; see Fig. 10.30.

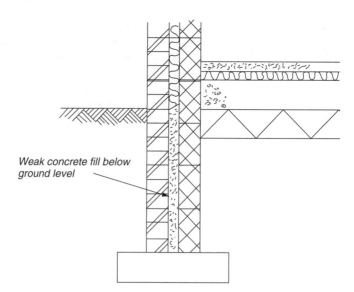

Figure 10.30 Cavity walls below ground level

Weak concrete fill below ground level

Brick cladding

With multi-storey buildings, the total loading of all the floors would be too much for normal 275 mm-wide cavity walling to support safely. In these situations, a permanent frame of reinforced concrete or structural steel is erected first, before external cavity walling is built around it; see Fig. 10.3.

Support for brick cladding

Although the bricklayer builds this cavity walling in exactly the same way as for low-rise structures, when used on multi-storey buildings, both leaves of brick masonry must be supported at every floor level. This ensures that permanent pressure does not build up on the brick masonry at ground level. Cavity wall cladding to multi-storey buildings is described as non-load bearing, because it does not carry the weight of floor slabs resting upon it.

Support for the outer skin of brick cladding using reinforced concrete edge beams, as shown in Figs 10.31 and 10.32, has generally been replaced by stainless steel angles or brackets, as shown in Figs 10.33 to 10.36.

You will notice that Figs 10.31 to 10.36 inclusive show a soft-joint filler of polyethylene foam strip under the supporting steel angle or concrete toe beams. This very important compression joint must be provided under each supporting angle of stainless steel, or reinforced concrete toe beam, around the external perimeter of a building at every floor level.

The horizontal compression joint, completely clear of mortar, permits different rates of expansion and contraction (up or down) to take place without restriction, between the building and the brick masonry cladding; see Fig. 10.36.

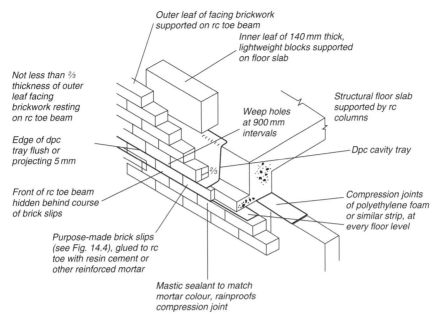

Outer leaf of facing brickwork
supported on rc toe beam

Inner leaf of 140 mm thick,
lightweight blocks supported
on floor slab

Not less than ⅔
thickness of outer
leaf facing
brickwork resting
on rc toe beam

Weep holes
at 900 mm
intervals

Structural floor slab
supported by rc
columns

Edge of dpc
tray flush or
projecting 5 mm

Dpc cavity tray

⅔

Front of rc toe beam
hidden behind course
of brick slips

Compression joints
of polyethylene foam
or similar strip, at
every floor level

Purpose-made brick slips
(see Fig. 14.4), glued to rc
toe with resin cement or
other reinforced mortar

Mastic sealant to match
mortar colour, rainproofs
compression joint

Figure 10.31 Support of cavity wall brick cladding on reinforced concrete toe beam

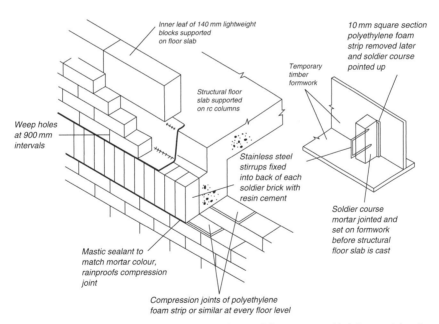

Inner leaf of 140 mm lightweight
blocks supported on floor slab

10 mm square section
polyethylene foam
strip removed later
and soldier course
pointed up

Temporary
timber
formwork

Structural floor
slab supported
on rc columns

Weep holes
at 900 mm
intervals

Stainless steel
stirrups fixed
into back of each
soldier brick with
resin cement

Soldier course
mortar jointed and
set on formwork
before structural
floor slab is cast

Mastic sealant to
match mortar colour,
rainproofs compression
joint

Compression joints of polyethylene
foam strip or similar at every floor level

Figure 10.32 Outer leaf of cavity wall facework supported on soldier course of bricks, cast in situ with reinforced concrete structural floor slab and edge beam

Unless these horizontal compression joints are provided at every floor level, serious cracking will result when the outer brickwork tries to expand and/or the load bearing structural frame shrinks slightly over the years.

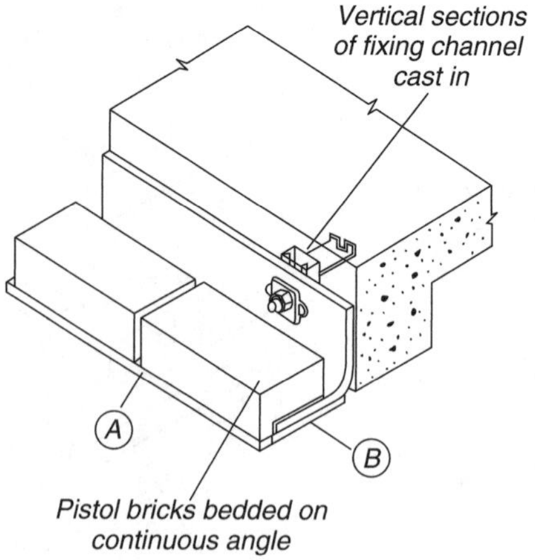

Figure 10.33 Heavy section stainless steel angle for continuous support of brick cladding

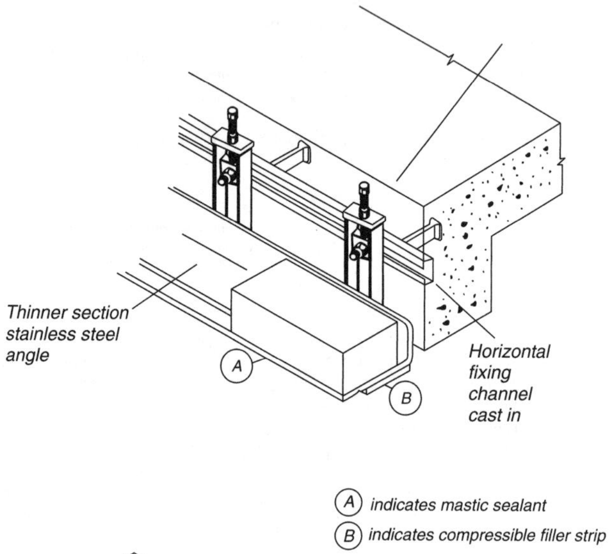

Vertical sections of fixing channel cast in

Pistol bricks bedded on continuous angle

Ⓐ   Ⓑ

Figure 10.34 Thinner section stainless steel angle with adjustable support brackets welded on at 450 mm intervals

Thinner section stainless steel angle

Horizontal fixing channel cast in

Ⓐ   Ⓑ

Ⓐ *indicates mastic sealant*
Ⓑ *indicates compressible filler strip*

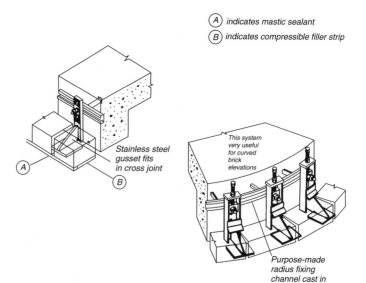

Stainless steel gusset fits in cross joint

Ⓐ   Ⓑ

This system very useful for curved brick elevations

Purpose-made radius fixing channel cast in

Figure 10.35 Separate stainless steel brackets supporting each end of stretchers

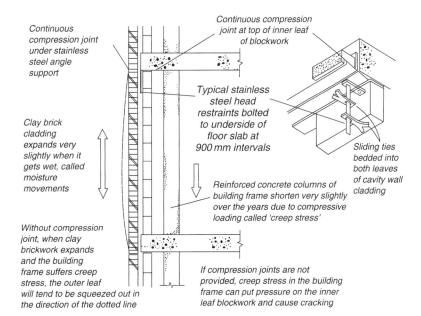

Continuous compression joint under stainless steel angle support

Continuous compression joint at top of inner leaf of blockwork

Typical stainless steel head restraints bolted to underside of floor slab at 900 mm intervals

Clay brick cladding expands very slightly when it gets wet, called moisture movements

Sliding ties bedded into both leaves of cavity wall cladding

Reinforced concrete columns of building frame shorten very slightly over the years due to compressive loading called 'creep stress'

Without compression joint, when clay brickwork expands and the building frame suffers creep stress, the outer leaf will tend to be squeezed out in the direction of the dotted line

If compression joints are not provided, creep stress in the building frame can put pressure on the inner leaf blockwork and cause cracking

Figure 10.36 The importance of compression joints in brick cladding to framed structures

Pistol bricks

In order to disguise the presence of the supporting steel angle, within the thickness of a mortar bed joint, the first course of bricks may be rebated, as shown in Fig. 10.37. These 'pistol bricks' can be produced as bricks of special shape if the manufacturer is given sufficient notice. Alternatively they can be cut on a masonry bench saw.

Before the importance of horizontal compression joints at every floor level of multi-storey buildings was fully realised, great problems developed from the use of brick 'slips', which were used to disguise concrete toe beams. For these reasons, stainless steel angles or brackets are currently preferred for the support of brick cladding; see Figs 10.33 to 10.35.

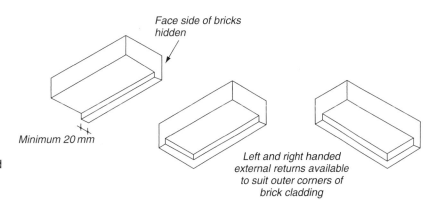

Face side of bricks hidden

Minimum 20 mm

Left and right handed external returns available to suit outer corners of brick cladding

Figure 10.37 Pistol bricks used to hide thickness of heavy section support angles

Tying back

In addition to providing support at every floor level for panels of external brick cladding, they must be securely tied back to the structural frame of a building. The non-corroding metal ties for this are used in addition to the normal cross-cavity wall ties, required between inner and outer skins of masonry, as shown in Fig. 10.13.

Vertical movement joints

The outer leaf of face brickwork in cavity wall cladding to a framed building will expand and contract due to changes in temperature and moisture content.

The horizontal compression joints shown in Figs 10.31 to 10.35 allow up and down movements in the brickwork to take place freely.

Allowance must also be made for sideways expansion and contraction, however, so that the outer leaf of brickwork does not cause damage to itself or the structure.

A typical vertical movement joint, with expanded plastic foam joint filler projecting, can be seen in Fig. 10.38. This shows 102 mm thick brick cladding.

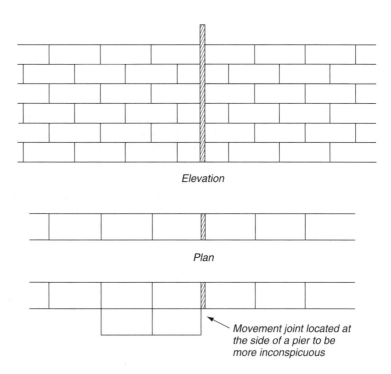

Elevation

Plan

Movement joint located at the side of a pier to be more inconspicuous

Figure 10.38 Brick cladding showing the position of a vertical movement joint

Vertical movement joints should be provided at intervals: between 9 and 12 m for clay brickwork; not exceeding 7 to 9 m for sand lime (calcium silicate) bricks; and at 6 m intervals for concrete blockwork.

Plainly, vertical movement joints form a 'straight joint' weakness in a wall, and so must be strengthened with stainless steel slip-ties every fourth course – as shown in Chapter 13 (Fig. 13.10).

Head restraints

When a building has load bearing walls (see Fig. 10.2), the top of these walls are rigidly held in position by the weight of floors bearing upon them.

The external brick cladding, and internal walls within framed structures, have soft compression joints at the underside of floor slabs; see Figs 10.31 to 10.35. These must be held in position by some form of head restraint. The inset to Fig. 10.36 shows a typical example of a stainless steel sliding head restraint.

11 Damp prevention

It is important that the bricklayer should understand the causes of dampness and the methods of damp prevention in the construction of a brick wall. Dampness may occur even after precautions have been taken, if precautions are based on insufficient knowledge resulting in lack of care in the application of damp prevention.

Forming a damp proof course (dpc)

The spread of dampness into the main structure of a building is prevented by a damp proof course (or damp course). This is a layer of non-absorbent material, examples of which are given in Table 11.1. The damp proof course must remain intact. All flexible dpc materials must be laid upon a thinly spread bed of fresh mortar, to provide full support across frogs, perforations and joints between bricks. Laying the bed for the next course of bricks ensures that the dpc is then neatly sandwiched and protected. It must not be broken by being laid on an uneven surface, by pebbles in the mortar if a slate damp course, or by clumsiness on the part of the bricklayer in dropping bricks on to the newly laid damp course.

Routes for damp penetration of buildings

Damp affects walls in several ways:

1. Moisture rising up from the ground
2. Penetration through the face of walls, caused by driving rain
3. Moisture percolating downwards from the top of walls or chimney stacks.

Table 11.1 Types of damp proof course

General classification	Description	Type of material	Comments
Flexible	Rolled materials available in a range of widths to suit different wall thicknesses and for folding into cavity trays. Provide the most economical and convenient type of dpc.	Hessian-reinforced bitumen Ditto with thin lead foil sandwiched within Fibre reinforced bitumen Ditto with lead foil	Standard dpc material available in rolls of approx. 8 m in length. Must be minimum 100 mm running laps, and at corners or T-junctions. Must be laid upon a thin bed of fresh mortar. Always store rolls on end to avoid distortion and cracking. Keep this type of dpc in a warm place before use, to make unrolling easier in cold weather.
		Pitch-polymer Black polythene	Thinner, tougher dpc which will not squeeze out under long term pressure from brickwork. Easier to use where projecting vertical dpc must be folded against sides of window and door frames. Available in 24 m length rolls. Same minimum laps as above.
	Metal	Sheet lead Sheet copper	Lead and copper dpc should be painted both sides with bituminous paint before use, to prevent possible corrosion from cement, lime or soluble salts in clay bricks. Laps a minimum of 100 mm, or joints 'welted', particularly if used to resist downwards penetration of dampness. Must be laid upon a thin bed of fresh mortar.
Semi-rigid		Natural mined rock asphalt, or a carefully controlled mixture of bitumen and limestone aggregate.	Heavy blocks of material melted down on site and spread in two or three separate coats while molten. Normally only used as a wall dpc if mastic asphalt tanking to a basement is being applied in the building (see Figs 11.12 and 11.13).
Rigid	Engineering bricks, minimum two courses bedded in Group (i) cement mortar 1:3 Only used for horizontal dpc.	Black or red Class 'A' engineering bricks, with a water absorption not exceeding 4.5%.	A very good method of creating a dpc in a free standing boundary wall, for reasons given in Chapter 14 and shown in Fig. 14.9. Cross joints may be left 'open', to prevent dampness rising up through the mortar. Engineering bricks are not suitable to resist downwards penetration of dampness.
	Slates, two courses bedded in Group (i) cement mortar 1:3	Sound, hard Welsh slate only. Soft flaky slates are not suitable for a dpc.	A traditional dpc, with the slates bedded 'stretcher bond', which is usually described in Bills of Quantities as 'laid breaking joint' (see Fig. 11.18). Not used in modern construction due to cost, and also because the slightest building settlement will crack the slates. Overall thickness of a slate dpc is 35 to 40 mm.

In building up any wall construction, and having in mind the causes of dampness, the passage of the moisture should be followed and the necessary precautionary measures should be taken in their logical sequence to prevent it entering the interior of the building; see Fig. 11.1.

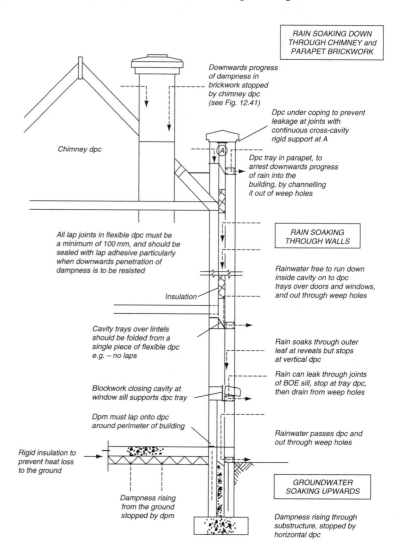

RAIN SOAKING DOWN
THROUGH CHIMNEY and
PARAPET BRICKWORK

Downwards progress
of dampness in
brickwork stopped
by chimney dpc
(see Fig. 12.41)

Dpc under coping to prevent
leakage at joints with
continuous cross-cavity
rigid support at A

Chimney dpc

Dpc tray in parapet, to
arrest downwards progress
of rain into the
building, by channelling
it out of weep holes

RAIN SOAKING
THROUGH WALLS

All lap joints in flexible dpc must be
a minimum of 100 mm, and should be
sealed with lap adhesive particularly
when downwards penetration of
dampness is to be resisted

Rainwater free to run down
inside cavity on to dpc
trays over doors and windows,
and out through weep holes

Insulation

Rain soaks through outer
leaf at reveals but stops
at vertical dpc

Cavity trays over lintels
should be folded from a
single piece of flexible dpc
e.g. – no laps

Rain can leak through joints
of BOE sill, stop at tray dpc,
then drain from weep holes

Blockwork closing cavity at
window sill supports dpc tray

Dpm must lap onto dpc
around perimeter of building

Rainwater passes dpc and
out through weep holes

Rigid insulation to
prevent heat loss
to the ground

Dampness rising
from the ground
stopped by dpm

GROUNDWATER
SOAKING UPWARDS

Dampness rising through
substructure, stopped by
horizontal dpc

Figure 11.1 Routes for dampness to soak into a building, and dpc provision to resist it

1. Moisture rising up from the ground

If soil investigation tests on a site indicate that the ground waters contain soluble sulphates, then the cement used in sub-structure brickwork and foundation concrete must be sulphate resisting.

The construction at the wall base, in conjunction with the floor construction, is shown in the following order:

1. Timber or suspended floors above ground level
2. Solid concrete floors above ground level
3. Concrete floors below ground (basement floors).

Timber floor above ground level

Location of main horizontal dpc

All timber wall plates, and structural floor joists associated with timber flooring, must be pressure impregnated against fungus and beetle attack.

(a) Moisture rises through the foundation concrete and also penetrates the wall in contact with the soil. To prevent it continuing up the wall, some impervious material must be built into the wall; this is called the horizontal dpc, which should be a minimum of 150 mm above ground level (Fig. 11.2).

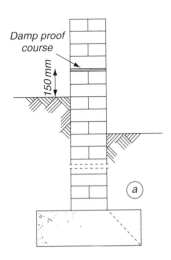

Figure 11.2 Horizontal dpc provision in free-standing wall

Solid walls

A one-brick boundary wall is shown with horizontal dpc positioned 150 mm above ground level. This height is used to prevent splashing of rainwater over the dpc level. Solid brick external walls may be found on existing old properties when renovation work is carried out. (The apprentice should note that this particular wall thickness is unsuitable with regard to weather penetration and thermal insulation, and would therefore require additional constructional treatment to outer and inner faces.)

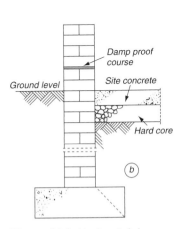

Figure 11.3 Horizontal dpc provision in external wall of building

Sealing the ground under a building

(b) Some provision must also be made to prevent damp air rising within the building, carrying with it the odours associated with decaying vegetable and animal matter. This is done by covering the whole of the site within the walls of the building with a layer of concrete from 100 mm to 150 mm thick; this is called the 'site concrete' or 'oversite' (Fig. 11.3). The oversite must be positioned so that its top surface is not below the highest level of the ground surface or paving around the building.

Underfloor ventilation against dry rot

(c) Having placed the oversite, the floor must be supported, so as to allow free passage of air – to prevent the floor timbers from rotting. This is achieved by building honeycomb sleeper walls. Two types are illustrated in Fig. 11.6. The sleeper walls nearest to the main wall are positioned as in Fig. 11.4; it provides freedom of movement when the bricklayer is building the wall and less possibility of mortar droppings collecting between the main wall and the sleeper wall. Figure 11.5 shows a construction that will provide efficient support and very important ventilation to the floor timbers.

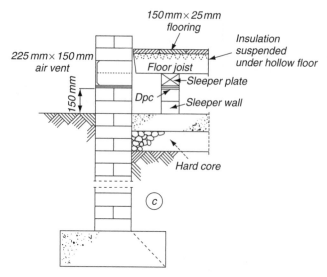

Figure 11.4 Solid wall – location of sleeper wall parallel to external wall

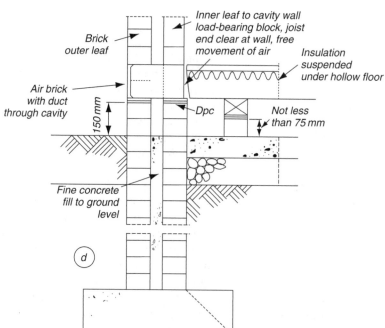

Figure 11.5 Cavity wall – minimum height of sleeper wall which will permit adequate underfloor ventilation

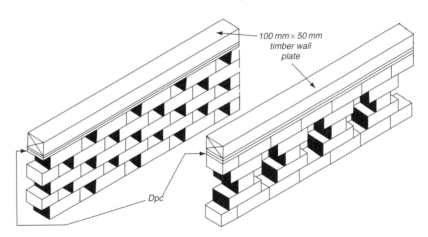

Figure 11.6 Honeycomb sleeper walls

Damp, stagnant air provides ideal growing conditions for a fungus commonly called 'dry rot' to take root in constructional timbers. Although needing moisture to grow, the fungus roots destroy the timber, leaving it split, broken and 'dry' thereby explaining the name.

Hollow floors have to be insulated to prevent cold air rising into the building (Figs 11.4 and 11.5).

Bedding wall plates

(d) Having taken the necessary precautions against dampness, it is the bricklayer's job to bed the wall plates in readiness for the fixing by the carpenter of joists and flooring. The wall plate, usually of 100 mm × 50 mm or 100 mm × 75 mm sawn timber, is placed immediately above damp course level. This completes the construction so far as the bricklayer is concerned, with the exception of the building-in of air bricks, which must be placed in suitable positions; their function is to ensure complete under-floor ventilation. It may happen that a hollow timber floor of one room is adjacent to a solid concrete floor in the next room, thus presenting some difficulty with regard to through ventilation. This may be overcome by inserting a series of drainpipes under the solid floor (Fig. 11.7), thus allowing a satisfactory flow of air to be maintained from front to back with terraced houses particularly.

It will be noted in the illustration that a difference in floor levels occurs. Where the adjacent floor levels are the same, thickening of the solid concrete floor, as shown in Fig. 11.8, will ensure that any joist timbers touching the partition wall are protected from the effects of dampness. This is achieved by similar linking of dpm and dpc. An alternative construction is the application of a waterproof membrane or vertical damp course (Fig. 11.9).

The size of a ground-floor joist in ordinary domestic construction is usually 125 mm × 50 mm laid on edge, with sleeper walls placed at 1350 to 1500 mm intervals to give satisfactory support. An alternative type of suspended ground-floor construction, using pre-cast concrete beams and blocks, is shown in Chapter 10, Fig. 10.8.

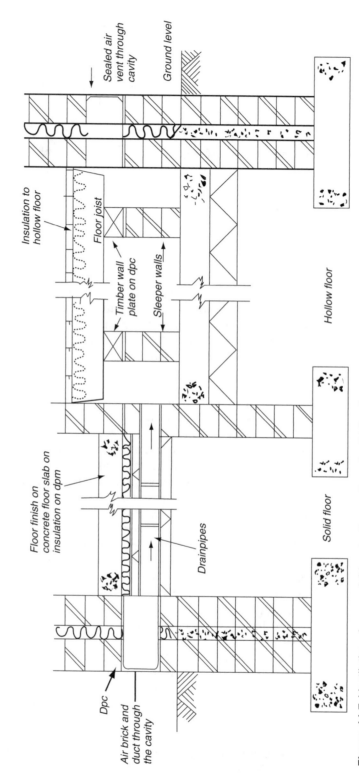

Figure 11.7 Ventilation pipes are laid under the section of solid flooring, to ensure vigorous through-ventilation

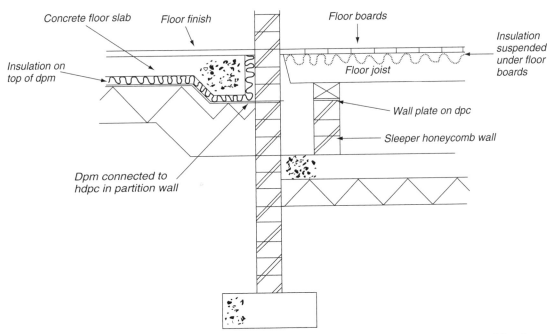

Figure 11.8 Protecting floor timbers from dampness at the junction of solid and suspended ground flooring

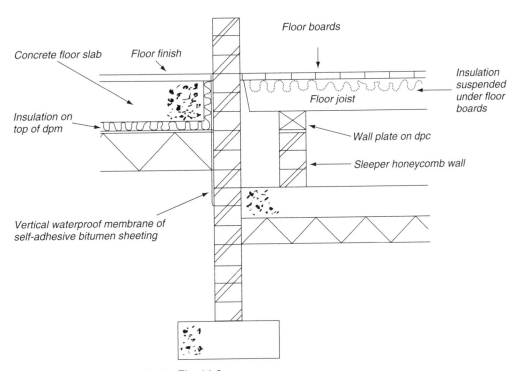

Figure 11.9 Alternative construction to Fig. 11.8

Solid concrete floors above ground level

The sequence of damp penetration can be followed as before. The horizontal damp proof course level must never be above the floor level. Brick rubble or hardcore laid directly beneath the concrete floor will not only prevent settlement, but, being of a porous nature, will help prevent dampness. Its thickness should be approximately that of the concrete floor (Fig. 11.10).

Solid ground floors have to be insulated, according to the current Building Regulations, to provide resistance to unacceptable heat loss through the floor. This can be achieved in various ways but the most common is to place the insulation on top of the damp proof membrane which is placed on a blinding layer on top of the hardcore.

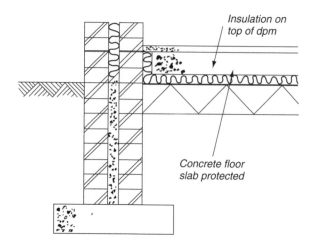

Insulation on top of dpm

Concrete floor slab protected

Figure 11.10 Damp proofing a solid ground floor slab with the dpm under the concrete floor slab

This method of construction protects the concrete floor slab from any moisture or harmful salts. The only problem with this method is the risk of damage to the dpm and insulation when laying the concrete floor slab; see Fig. 11.10.

An alternative method is shown at Fig 11.11. This method is easier but the concrete floor slab is not protected against the ingress of moisture or harmful salts.

Basement floors below ground level

When a building has habitable rooms below ground level they have to be constructed not only to act as retaining walls but also have to offer resistance to moisture from the ground to the inside of the building or to any material used in the construction that would be affected by moisture.

Moisture penetrates through the walls in contact with the surrounding soil and upwards through the foundations and the whole floor of the

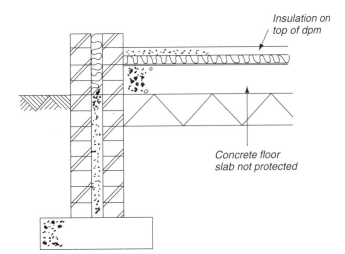

Insulation on top of dpm

Concrete floor slab not protected

Figure 11.11 An alternative method with the dpm on top of the concrete floor slab

building. To resist this moisture penetration, the base of a structure must be damp proofed throughout the floors and walls below ground level and to at least 150 mm above ground level. This is called 'tanking'.

Basement tanking

Mastic asphalt, applied molten in three separate coats, is one way of reliably resisting groundwater pressure. Such as when the general groundwater pressure level in the surrounding soil is higher than the basement floor level.

Alternatively, self-adhesive bituminous sheeting applied to floor and walls, carefully lapped and sealed at joints, may be considered sufficient for damp proofing a basement on a well-drained site.

Both methods of tanking a basement need permanent protection. A 'loading coat' of in-situ reinforced concrete is necessary on basement floors to 'hold down' and protect mastic asphalt or bituminous sheeting. Protection with brickwork is required to prevent damage, and to hold back vertical tanking on basement walls.

Tanking from outside
Figure 11.12 shows tanking which has been applied externally. This is the best method of working, because eventually groundwater pressure will tend to keep the vertical waterproofing pressed back in place.

The last operation is to build an external brick wall covering, one half brick in thickness, to protect the tanking when the soil is back-filled around the building.

Tanking from inside
Figure 11.13 shows the opposite way of working, that is tanking applied from inside: the method likely to be used where an existing basement is to be waterproofed. The same concrete floor loading slab and protection to vertical tanking are necessary. Care should be taken beforehand to check the reduction in basement headroom that this loading slab thickness will cause.

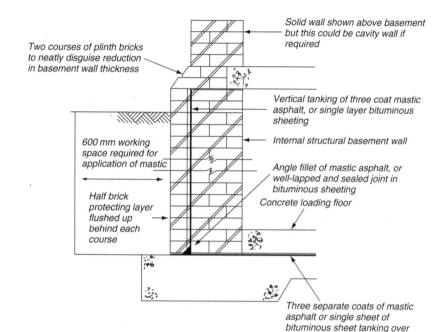

Two courses of plinth bricks to neatly disguise reduction in basement wall thickness

Solid wall shown above basement but this could be cavity wall if required

Vertical tanking of three coat mastic asphalt, or single layer bituminous sheeting

Internal structural basement wall

600 mm working space required for application of mastic

Angle fillet of mastic asphalt, or well-lapped and sealed joint in bituminous sheeting

Concrete loading floor

Half brick protecting layer flushed up behind each course

Three separate coats of mastic asphalt or single sheet of bituminous sheet tanking over concrete foundation

Figure 11.12 Basement tanking – external application

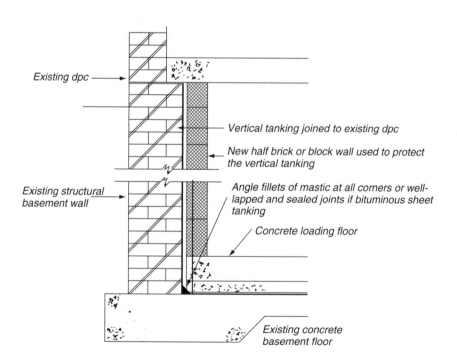

Existing dpc

Vertical tanking joined to existing dpc

New half brick or block wall used to protect the vertical tanking

Existing structural basement wall

Angle fillets of mastic at all corners or well-lapped and sealed joints if bituminous sheet tanking

Concrete loading floor

Existing concrete basement floor

Figure 11.13 Basement tanking – inside application

2. Moisture entering through the face of walls

Some consideration has already been given to this in respect of brick walls below ground level, where vertical damp proofing is necessary. In walls above ground level exposed to the atmosphere, vertical damp proofing is, in the true sense of the term, unnecessary.

Cavity wall construction, dealt with in detail in Chapter 10, is the standard form of building the external walls of domestic and commercial structures of brickwork and blockwork. The 50 to 75 mm-wide air space acts as a vertical damp proof 'zone' within all the external walls of a building. The cavity is a complete break between the outer leaf which can become saturated and the internal wall surface.

Solid walls

One-brick thick solid walls, if properly built and not exposed to driving rain, will resist the passage of moisture to some degree.

If moisture penetrates a brick wall, it will probably occur where the mortar joint is in contact with the brick (the 'interface'), because of insufficient bond between them. For instance, where a very hard type of mortar is used there is a tendency for the mortar to shrink away from the bricks, leaving hairline cracks through which moisture can pass.

For this reason, one-brick solid walling is unsuitable, together with the fact that it allows heat energy to escape readily by conduction.

Chapter 10 (Fig. 10.6) shows how the insulated cavity space reduces heat loss and provides a barrier to the cross-cavity penetration of dampness. It should be noted that the solid one-brick thick external walling of existing buildings and maintenance of other brickwork generally is raked out and re-pointed, to improve resistance to the weather as well as its stability.

Excessively strong, cement-rich mortar should be avoided when re-pointing brickwork. A 1:1:5 mix is quite strong enough for the majority of facing bricks, in order to allow the natural weathering processes of wetting and drying, absorption and evaporation to take place evenly, from brick and joint surfaces respectively.

Cavity walls

With those dual requirements of the Building Regulations, that external walls of dwellings shall resist rain penetration and conserve heat energy, standard cavity wall construction as shown in Chapter 10 (Fig. 10.1) is the most economical form of construction for external walling.

Where a cavity is closed at the reveals of windows and doors, a vertical dpc separates inner and outer brick masonry to prevent rain from soaking across.

Where the cavity is 'bridged' by lintels across openings, a 'gutter' or 'tray' of flexible dpc material is formed to gather up any rainwater that has soaked in and direct it out again via weep holes as shown in Fig. 11.1.

Trays of flexible dpc under brick and tile sills similarly prevent rain leakage at joints from progressing to the internal wall surface. Cavity walling has been examined in general terms, in this present chapter, from a number of directions. Cavity wall constructional details are dealt with in greater depth in Chapter 10.

3. Moisture percolating through the tops of walls

Walls below the roof level of a building get wetted by rain on one side only. The brickwork below the overhanging eaves of a pitched roof, an example is shown in Chapter 10 (Fig. 10.2), is very nicely protected at the top.

Parapet walls

These are brick walls that stand up around the edges of a roof. They provide somewhere against which to seal the roof covering of asphalt or sheet material, and can also act as a permanent barrier around a roof.

However, the brickwork of a parapet wall, as illustrated in Fig. 11.1 and Figs 11.14 and 11.15, is exposed to the weather on both sides and the top. For this reason great care must be taken to prevent dampness entering the walls of buildings in this way.

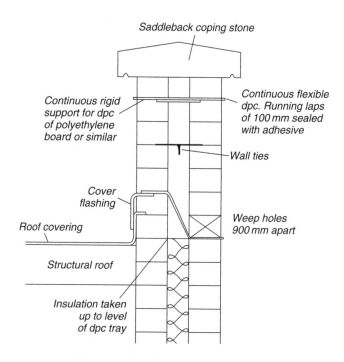

Figure 11.14 Parapet cavity tray tilted away from roof

Parapet wall construction
Many parapet walls have been built as one-brick or $1\frac{1}{2}$ brick thick solid walling, but it has been found that cavity wall construction for parapets is preferable in preventing downwards penetration of dampness.

A dpc tray incorporated across the cavity wall width collects any rainwater that enters the cavity from either direction and disperses it via weep holes; see Fig. 11.14.

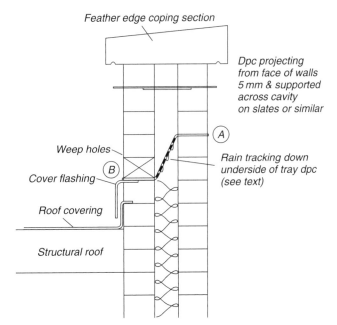

Feather edge coping section

Dpc projecting
from face of walls
5 mm & supported
across cavity
on slates or similar

(A)

Weep holes

(B)

Cover flashing

Rain tracking down
underside of tray dpc
(see text)

Roof covering

Structural roof

Figure 11.15 Parapet cavity tray tilted towards roof, showing risk of rain penetration at A and B, channelling rain on to the inner leaf in areas of high exposure

Purpose-made nylon fibre plugs can be put into these weep holes to prevent any water staining of the facework. Tilting the cavity tray the opposite way, towards the roof, as shown in Fig. 11.15, runs the risk of directing dampness on to the inner leaf of masonry.

Chimney stacks

This brickwork is open to the elements on four sides as well as the top, and also suffers the stress of thermal expansion and contraction due to the passage of hot gases within.

For these reasons, it is not surprising that chimney stack brickwork is generally the first that needs re-pointing or some other maintenance after only 20 years into the life of the building. Choice or specification of bricks and mortar are therefore particularly important for chimney stack brickwork.

Choice of bricks

Facing bricks in the exposed locations of parapet walling and chimney stacks need to be resistant to frost. This requirement obviously influences the choice of facing bricks for the whole building, because parapet and chimney stack brickwork must match colour and type of the rest of the facework.

Choice of mortar

Reference should be made to Chapter 2 (Table 2.8) for a mortar strength also suited to these exposed positions. The use of sulphate resisting cement in the bricklaying mortar for parapets and chimneys may be advisable if the clay bricks to be used contain soluble salts. This to avoid long-term deterioration of this exposed brickwork by sulphate attack if ordinary Portland cement were to be used.

Weathering to the top of walling

Some form of projecting coping is preferable to protect the top of walls, to function rather like an umbrella.

The features of a good coping for a cavity wall parapet around a building, shown in Fig. 11.16, are similar to those detailed in Chapter 13 for solid boundary walling.

Pre-cast concrete or natural stone copings with the features illustrated in Fig. 11.16 provide the best long-term protection to parapet brickwork.

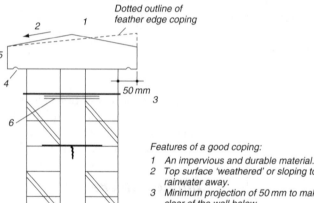

Dotted outline of
feather edge coping

50 mm

Features of a good coping:

1 An impervious and durable material.
2 Top surface 'weathered' or sloping to tip
   rainwater away.
3 Minimum projection of 50 mm to make raindrops fall
   clear of the wall below.
4 Continuous groove or drip throating to prevent
   raindrops from tracking across on to wall face.
5 As few joints as possible between units of coping.
6 Continuous layer of dpc, supported across the cavity,
   either directly under the coping or one course down,
   to arrest leakage at joints.

Figure 11.16 Section through parapet wall with saddleback coping

Always ensure the copings are laid to ensure the eye line is correct. The eye line is determined by which edge is seen by the eye. As a general rule copings above head height have the eye line to the lower edge.

Similarly the type of chimney stack coping illustrated in Fig. 11.1 and also in Chapter 13 (Fig. 13.11) include the foregoing desirable features to ensure efficient long-term performance.

The variety of BOE cappings shown in Fig. 11.17 look more attractive, but have many more joints, and for this reason are less successful in preventing rainwater from saturating parapet brickwork and the tops of other free-standing walls.

A number of brick manufacturers have developed two-part brick systems of cappings and copings for walls, to improve the performance of the parapet dpc plus their solidity and appearance. For examples see Chapter 13, Fig. 13.14.

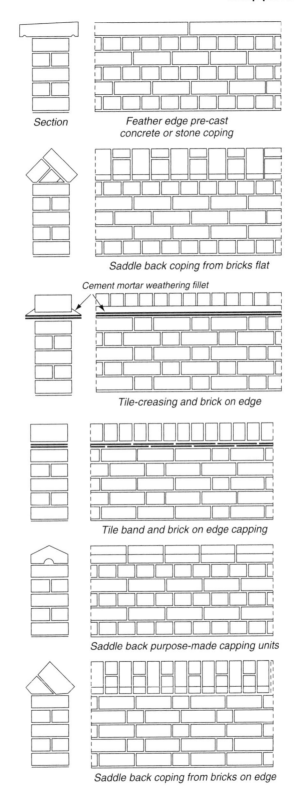

Section

Feather edge pre-cast
concrete or stone coping

Saddle back coping from bricks flat

Cement mortar weathering fillet

Tile-creasing and brick on edge

Tile band and brick on edge capping

Saddle back purpose-made capping units

Saddle back coping from bricks on edge

Figure 11.17 Finishes to tops
of walls

Types of damp proof course

The current building regulations state that no wall or pier shall permit the passage of moisture from the ground to the inner surface of a building.

Damp proof courses are classified under the following headings:

- Rigid
- Semi-rigid
- Flexible.

Rigid

Materials selected for damp proof courses must be capable of resisting the passage of moisture, and when applied to the wall must be continuous throughout. This continuity may be broken either by bad workmanship or, where a rigid type of damp proof course is used, e.g. slate, by settlement of the building. Failures will therefore be prevented by the selection of a suitable material to supplement good craftsmanship and a reliable foundation.

Slates must be hard and rough in texture and be laid in 3:1 cement mortar, in two bonded layers. See Table 11.1 and Fig. 11.18.

Bricks
Two courses of dense engineering bricks, laid bonded in cement mortar, act as a very effective damp proof course especially for boundary walls and piers; see Fig. 11.19.

Semi-rigid

Mastic asphalt
This is used nowadays mainly as a damp proof membrane. When used as a damp proof course it was spread hot in one or two coats to a thickness of approximately 13 mm and is impervious to moisture. Could be affected from moderate settlement in a wall which could cause cracking. It is an expensive form of a damp proof course; see Fig. 11.20.

Flexible

There are many materials which can be used as a damp proof course which are flexible.

Lead and copper
Lead and copper are very good but expensive. Lead is used mainly by the plumber for flashings and should weigh not less than 19.5 kg/m^2.

Lead and copper should be coated with bitumen paint to prevent any corrosion from lime mortar and should be laid in rolls the full thickness of the wall and lapped at joints and intersections at least 100 mm.

It is possible to have a dpc of bitumen with a thin layer of either lead or copper inside. All rolls of lead and copper should be stored on end to prevent distortion.

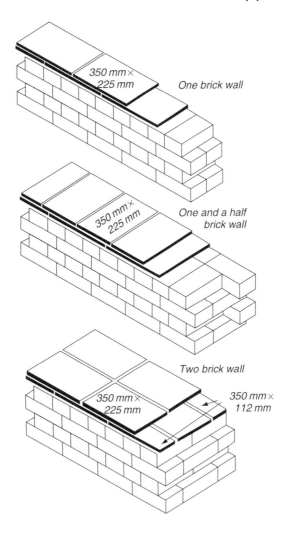

One brick wall

350 mm × 225 mm

One and a half brick wall

350 mm × 225 mm

Two brick wall

350 mm × 225 mm

350 mm × 112 mm

Figure 11.18 Slate damp proof courses

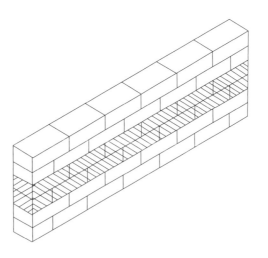

Figure 11.19 Engineering brick damp proof course

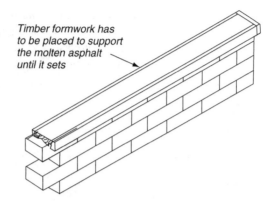

Timber formwork has to be placed to support the molten asphalt until it sets

Figure 11.20 Mastic asphalt damp proof course

Bitumen-based products

There are many types of bitumen dpcs and all are flexible but could squash under pressure. Bitumen dpcs should be used in rolls to suit the wall widths and bedded on mortar and lapped at joints and intersections by a minimum of 100 mm. All rolls should be stored on end to prevent distortion.

Polythene

Black low density polythene, like bituminised felt, should be laid in mortar with laps at least equal to the width of the dpc; see Fig. 11.21.

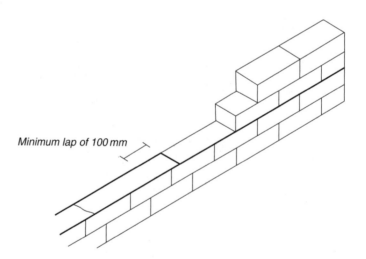

Minimum lap of 100 mm

Figure 11.21 Flexible damp proof courses

12 Chimneys, flues and fireplace construction

Many centuries ago when people first decided to bring the fire they used for cooking and heating inside their homes, the safest place for it was in the middle of the ground floor where everyone could keep an eye on it.

Smoke simply rose up into the rafters to escape through spaces in the roof covering.

Not until houses were built from non-combustible stone and brick was the fire placed against the wall between two attached piers called 'jambs' to form a fireplace recess.

Eventually the two jambs were joined with an arch and a chimney breast constructed above, leading to a hole in the wall behind for the smoke to escape.

Later, a vertical shaft was formed in the chimney breast to channel smoke up through the roof, inside a chimney stack of stone or brick, in order to protect the roof timbers from fire risk.

In order that the apprentice may clearly understand this important craft operation, this chapter has been divided into four main headings:

1. The types and positioning of flues and chimneys, as desired by the architect.
2. The Building Regulations controlling construction.
3. The construction of a chimney through the ground floor, first floor and roof of an ordinary domestic dwelling. By following the sequence of

operations, the reader should have no difficulty when confronted with a problem.

4. The fixing of fireplaces.

1. Types of chimney breasts and flues

Single fireplaces

The fireplace opening is formed on one side of the wall only, by the formation of attached piers called 'jambs'. Figures 12.1 and 12.2 illustrate the plans of various jamb arrangements, in solid walls, at the ground floor level or wherever the base or a chimney breast is commenced.

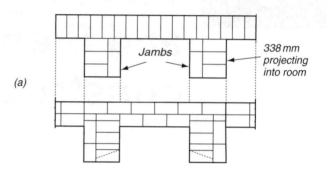

(a)

Jambs

338 mm projecting into room

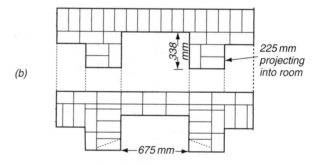

(b)

338 mm

225 mm projecting into room

675 mm

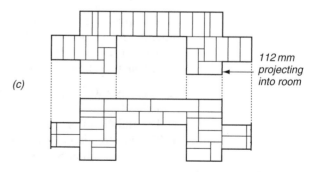

(c)

112 mm projecting into room

Figure 12.1 Single fireplaces, solid walls in English bond

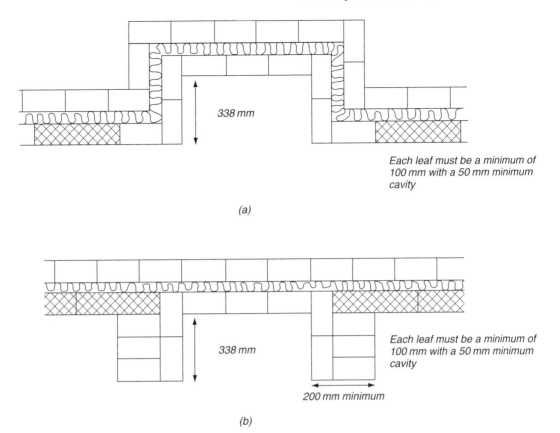

(a)

Each leaf must be a minimum of
100 mm with a 50 mm minimum
cavity

338 mm

Each leaf must be a minimum of
100 mm with a 50 mm minimum
cavity

338 mm

200 mm minimum

(b)

Figure 12.2 (a) Single fireplaces in cavity wall with external chimney breast. (b) Single fireplaces in cavity wall with internal chimney breast

Figure 12.1(a) shows a simple arrangement where the main wall forms the back of the fireplace opening. Figure 12.1(b) illustrates a similar example where the main wall is $1\frac{1}{2}$ bricks thick and the back of the fireplace is one brick thick. Figure 12.1(c) shows a typical example of a fireplace placed on external walls; the projection of the jambs into the room is lessened by the formation of an external breast or the breaking of the wall line on the opposite side of the wall. The advantage gained is the larger area available in the room in which the fireplace is situated.

Figures 12.2(a) and 12.2(b) show various details of chimney breasts in cavity walls with building blocks on the inner leaf.

Figure 12.3 shows a chimney breast constructed on an internal wall.

The width and height of the fireplace opening depends on the type of stove or grate to be inserted, while the width of the jambs depends upon the width of the chimney breast required on the upper floors.

The minimum depth of the fireplace opening is 338 mm, to allow a 225 mm flue to be formed together with a covering of 112.5 mm of brickwork when the chimney breast is being constructed above the lintel or arch level of the fireplace opening.

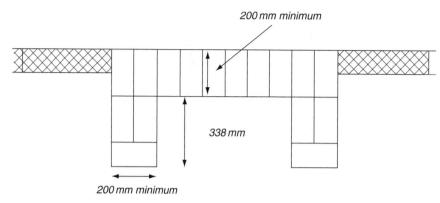

Figure 12.3 Single fireplace on an internal wall

Double or back-to-back fireplaces (Fig. 12.4)

This is a typical example of fireplace construction in the terrace or semi-detached type of dwelling. The fireplaces are formed on the party wall. Figure 12.5 shows a method of bonding adopted by some bricklayers to save the labour of cutting. This practice should be discouraged as bricks are often omitted where the straight joints occur and 'pockets' are formed, which are filled with rubbish – an example of bad workmanship.

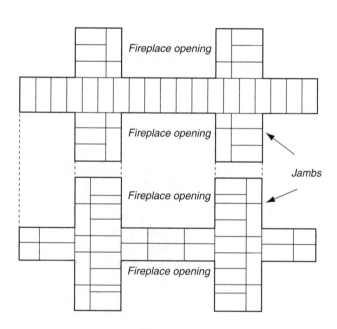

Figure 12.4 Back-to-back fireplaces

Interlacing fireplaces

These are fireplaces constructed on an internal wall and placed side by side; this arrangement lengthens the chimney breast, but the projection

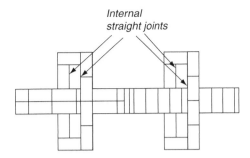

Figure 12.5 An example of bad bonding practice

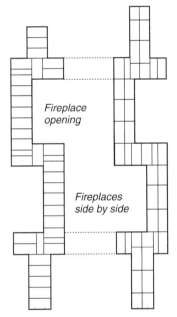

Figure 12.6 Interlacing fireplaces, ground floor

of the jambs into the room is reduced, giving greater room area (Fig. 12.6).

Angled fireplaces

Figure 12.7 shows the plans of the alternate courses of brickwork bonding at ground level. This is a difficult type of fireplace to construct because it entails a number of twists of the flue which are necessary to obtain correct positioning. Its construction will therefore not be shown at this stage.

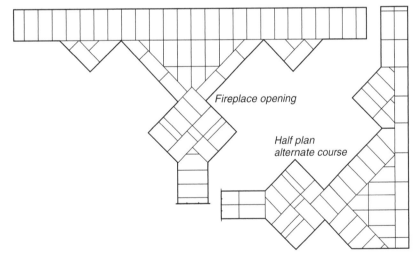

Figure 12.7 Plans of angle fireplace, English bond

Upper floor fireplaces

Figures 12.8 to 12.12 illustrate the bonding arrangements of upper floor fireplaces in the various types of construction. The alternate arrangement of flues in the interlacing fireplaces should be noted; this will be better understood when the grouping of fireplaces and chimneys has been considered.

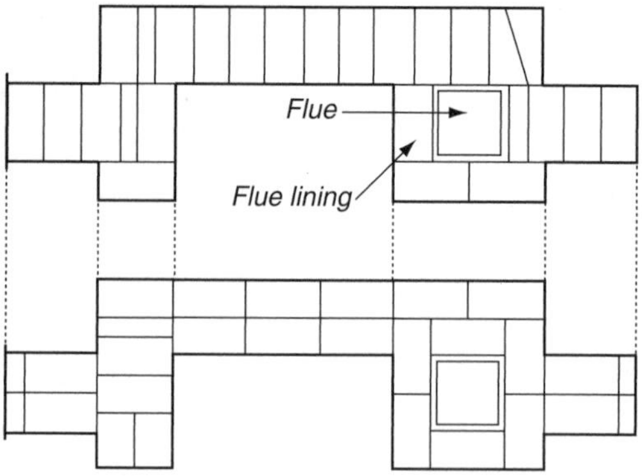

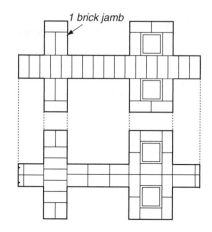

Figure 12.8 Single fireplaces, first floor

Figure 12.9 Back-to-back fireplaces, first floor

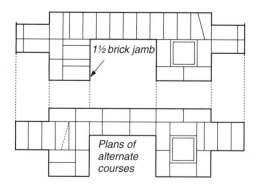

Figure 12.10 Single fireplace at first floor showing wider right-hand jamb

Grouping of fireplaces and chimneys

The details in the following section will mainly be found on existing buildings and have been included for those who are involved in the maintenance of such buildings.

In order to achieve sound construction and to reduce the number of chimney stacks emerging through the roof, fireplaces and chimneys were

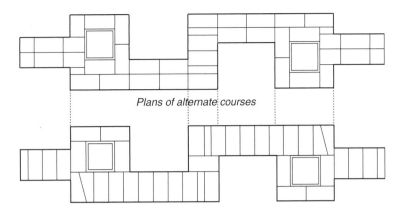

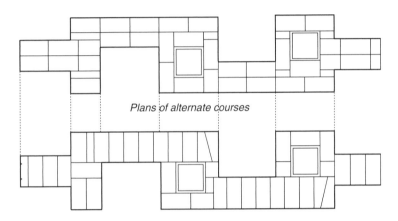

Plans of alternate courses

Figure 12.11 Interlacing fireplaces, upper floors

Plans of alternate courses

Figure 12.12 Interlacing fireplaces, upper floors, with central flue position

grouped to a central position. The fireplaces of each successive floor were positioned one above the other, and where possible the flues from fireplaces in adjacent rooms were gathered together before emerging through the roof. The base of a chimney breast was designed wide enough to support the fireplace construction that occurred in the floors above. Where buildings extended to four or five floors this is not always practicable, and in this case it was permissible to extend the upper floor chimney breasts in their length by corbelling. In simple domestic construction the latter point did not arise.

Figure 12.13 shows single breast fireplace construction on an external wall. The passage of the flues has been marked by dotted lines, and it will be noted that the flue from the ground floor fireplace has been gathered over to miss the first floor fireplace, while at roof level the flues from adjacent rooms have been grouped to form a single stack of four flues. The external chimney breasts have been connected by a face-brickwork semi-circular arch. The alternative is to continue the single chimney breast and to reduce it to stack size by means of circular ramps

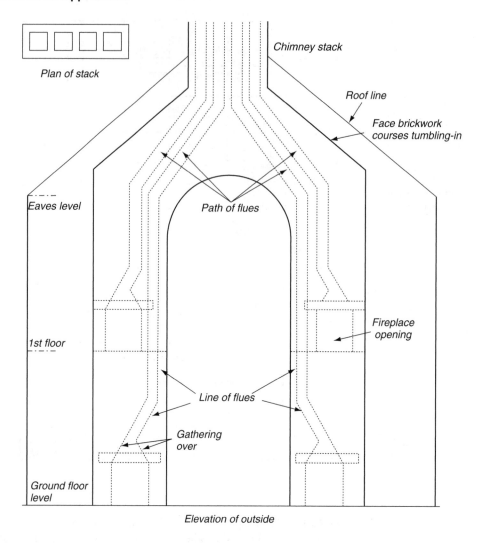

Plan of stack

Chimney stack

Roof line

Face brickwork
courses tumbling-in

Eaves level

Path of flues

Fireplace
opening

1st floor

Line of flues

Gathering
over

Ground floor
level

Elevation of outside

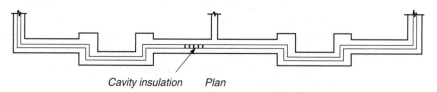

Cavity insulation   Plan

Figure 12.13 Single fireplaces on an external wall

(Fig. 12.14) or by tumbling (Fig. 12.13), where it will develop into a two-flued stack.

Figure 12.15 shows the grouping of chimneys and fireplaces on a double-breast party wall. The grouping of fireplaces from adjacent rooms takes place within the roof space and does not require the decorative

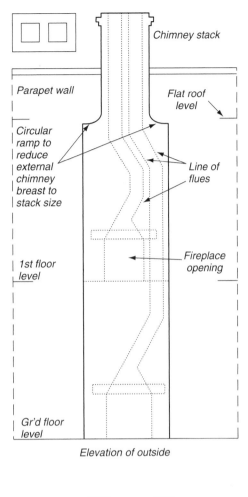

Chimney stack

Parapet wall

Flat roof level

Circular ramp to reduce external chimney breast to stack size

Line of flues

1st floor level

Fireplace opening

Gr'd floor level

Elevation of outside

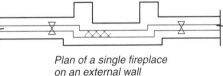

Plan of a single fireplace on an external wall

Figure 12.14 External chimney breast reduced to stack width using brick ramps

finish as in the external chimney breast, illustrated in Fig. 12.14, which accounts for the set-off a few courses above first floor ceiling level. The chimney stack emerging from the centre of the roof consists of eight flues, but the chimney breasts can be carried up individually, thus emerging one on each side of the ridge of the roof as a four-flue stack.

Figure 12.16 shows the grouping of interlacing fireplaces. In the description of types of fireplaces two methods of arranging the upper floor fireplaces were shown; both have their merits. Figure 12.12 is considered to be more straightforward, while Fig. 12.11 gives symmetrical planning. The chimney stack consists of eight flues in a line.

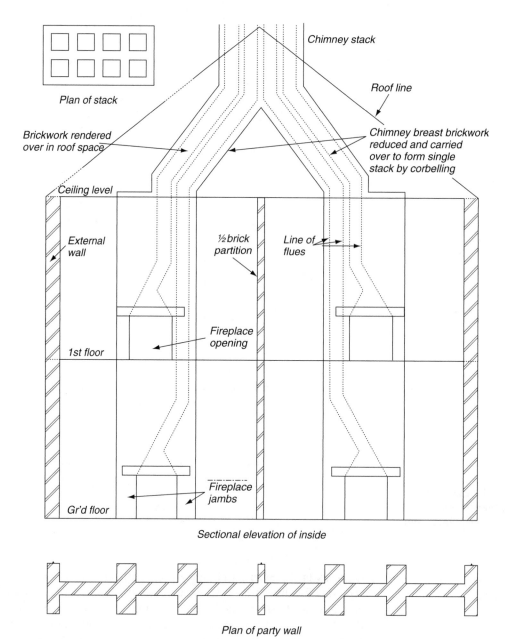

Plan of stack

Chimney stack

Roof line

Brickwork rendered over in roof space

Chimney breast brickwork reduced and carried over to form single stack by corbelling

Ceiling level

External wall

½ brick partition

Line of flues

Fireplace opening

1st floor

Fireplace jambs

Gr'd floor

Sectional elevation of inside

Plan of party wall

Figure 12.15 Back-to-back fireplaces on a party wall

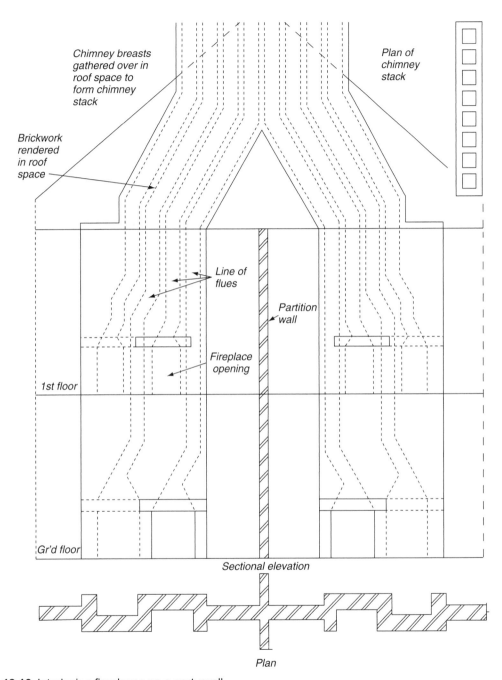

Figure 12.16 Interlacing fireplaces on a party wall

2. Regulations controlling the construction of chimney breasts and stacks

The construction of chimney breasts and stacks is controlled by Part J of the current Building Regulations. It is vital that the highest standard of workmanship and the current interpretation of Building Regulations is maintained throughout the construction of chimney breasts and stacks. Failure to do so can have devastating results.

The Building Regulations are concerned with stability, fire hazards and the discharge of products of combustion to the outside air. For the purpose of relating text and diagrams, sections will be numbered from No. 1 onwards, so that they will differ from references in the Building Regulations.

These cover the construction of flues, chimneys, fireplace recesses and hearths, for the purpose of an ordinary domestic fireplace appliance which will not exceed an output rating of 45 kW and a resulting flue not exceeding approximately 225 mm × 225 mm. Other clauses modify construction where gas appliances are used. The following descriptions are therefore an interpretation of the regulations in this sense, and this knowledge is sufficient for the construction of ordinary domestic fireplaces.

A chimney may be defined as the solid material surrounding a flue, while the flue is a duct through which smoke and other products of combustion pass. A chimney is therefore a single flue surrounded by at least 100 mm of brickwork and a stack consists of two or more flues surrounded by the requisite amount of brickwork. If the 'withes' or 'midfeathers' (dividing walls between flues) of a stack are half brick in thickness this will meet the requirements.

A flue formed in a brick wall is normally constructed to the dimensions of 225 mm × 225 mm or one brick by one brick; the cross-sectional area is reduced by the application of purpose-made flue liners. The minimum size of the flue to suit a 45 kW closed appliance is 175 mm diameter.

In simple residential buildings the size of a flue used solely for the purpose of discharging the fumes of a gas appliance into the open air must not be less than 12 000 mm^2 in cross-sectional area if circular, and 16 500 mm^2 if rectangular. Proprietary purpose-made flue blocks are available for the construction of gas flues, which must have a minimum dimension on plan of 90 mm.

1. Every flue must be surrounded by at least 100 mm thickness of brick-work, properly bonded and exclusive of the thickness of the flue lining.

2. Every chimney or stack must be built to a height of at least one metre above the last point of contact where it emerges from the roof. In the case of a gas flue this height is modified, but in no case must a chimney or stack be built up to a height greater than $4\frac{1}{2}$ times its least width at the last point of leaving the roof, unless it is adequately supported against overturning (Figs 12.17 and 12.18). The Building Regulations state that the top of a chimney carried up through the ridge of a roof, or within 1000 mm of it, and which has a slope on both sides of not less than ten degrees with the horizontal must be at least 600 mm above the ridge, but in all other cases at least 1 metre measured from the highest point in the line of junction with the roof.

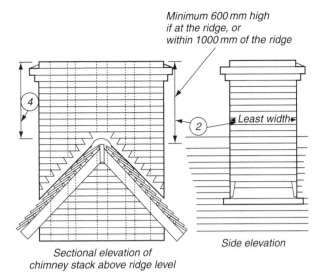

Minimum 600 mm high
if at the ridge, or
within 1000 mm of the ridge

Least width

Side elevation

Sectional elevation of
chimney stack above ridge level

Figure 12.17 Lead flashings to weatherproof junction between roof covering and a chimney stack that emerges through the ridge

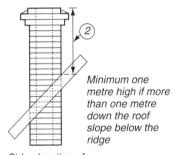

Minimum one
metre high if more
than one metre
down the roof
slope below the
ridge

Side elevation of
a chimney stack
emerging at mid roof

Figure 12.18 One metre minimum height where stack emerges mid roof

3. Every chimney must be built on suitable foundations approved by the local building control officer. In many traditional type houses chimney breasts have been projected from the main wall at an upper floor level; Fig. 12.19 illustrates a possible case in which the chimney breast of the ground floor is discontinued on the upper floor and in consequence the chimney has been corbelled-out on the external wall.

4. The 12 highest courses of every chimney or stack should be built in cement mortar (Fig. 12.17). A chimney stack passing through a roof must be properly protected against downwards penetration of dampness. This is achieved by building in a damp proof course within the brickwork.

5. Where chimney flues are inclined, the 'angle of travel' shall not be less than 60° to the horizontal. Chimney flues must be provided with a means of inspection and cleaning, via a gastight door set in a metal frame; see Fig. 12.20.

6. The inside of every chimney must be lined with non-combustible materials, e.g. fireclay or terracotta flue liners (Fig. 12.21); this work being carried out as the building of the chimney proceeds. The non-combustible linings are purpose-made units either square or circular in section to fit a one brick by one brick flue.

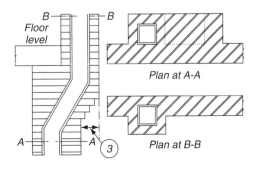

B — B
Floor
level

Plan at A-A

A — A

Plan at B-B

Figure 12.19 Corbelling at first floor level to form an external chimney breast

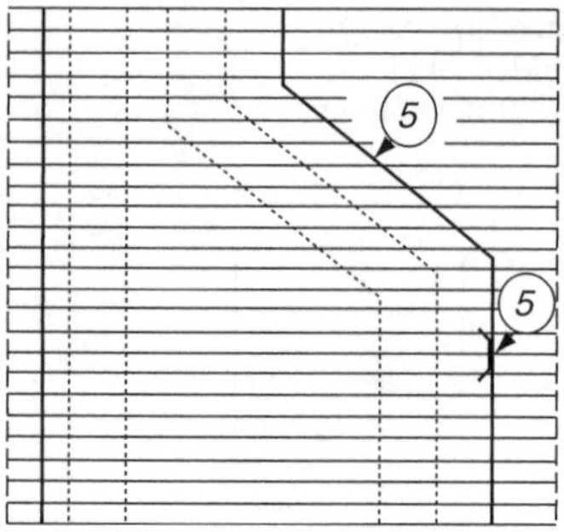

*Elevation of chimney
breast tumbled-in*

Figure 12.20 Travelling a flue in an external chimney

Flue liners must always be placed with sockets or rebates uppermost, to prevent leakage of any condensation which might form on the inside surfaces. Flue liners must also be jointed with fireproof mortar; see Fig. 12.34.

7. It is advisable that the outer surface of every chimney that is within a building or roof space is rendered up to the level of the outer surface of the roof or gutter. (Plastering to the internal walls of rooms largely takes care of this.) This rendering is for the additional protection of any combustible materials adjacent to the chimney where it passes through floors and roof.

(a) Woodwork must not be placed under any fireplace opening within 250 mm from the upper surface of the hearth except for the fillets or bearers supporting the hearth, or if there is an air space of not less than 50 mm between the underside of the hearth and any combustible material (see Figs 12.22 and 12.33).

(b) Woodwork must not be built into any wall nearer than 200 mm measured to the inside surface of a flue or to the inside of a fireplace recess (Fig. 12.21).

Figure 12.21 Plan of back-to-back fireplaces at first floor level (for numerical references, see text)

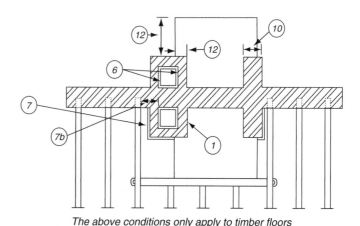

The above conditions only apply to timber floors

(c) Woodwork must not be placed closer than 40 mm to the surface of a chimney or fireplace recess, excepting floor boarding and skirting.

(d) If the surround of a fireplace opening is constructed of wood it should be at a distance of at least 150 mm measured horizontally and 300 mm measured vertically from the fireplace opening, and it must be backed with solid non-combustible material.

8. Where a chimney is adjacent to constructional steelwork or reinforced concrete, it is advisable to take precautions to ensure that the steelwork etc. is not affected by heat, with metal fixings at least 50 mm away.

9. Each fireplace must have its own flue taken to the outside air, with no branching of more than a single fireplace per flue.

10. The jambs of every fireplace opening must be at least 200 mm thick (Fig. 12.21).

11. The back of every fireplace opening on a party wall or party structure must be at least 200 mm thick.

This 200 mm thickness of brick masonry is also necessary to separate flues on either side of a party wall up to the level of the top floor ceiling. Reduction to 100 mm thickness between flues takes place at this point, where the bulk of the chimney breast is discontinued and the flues are grouped into one stack for passing through the roof covering (see Fig. 12.32).

It is desirable to keep the air in a flue warmer than the outside atmosphere, so that the air in a flue is rising continuously; the warmer the air in a flue, the quicker is the flow of air, and with it the gases of combustion, while the colder the air, the slower will be the flow, a condition which can lead to 'downdraught' and smoky chimneys.

12. The constructional concrete hearth must extend at least 150 mm beyond each side of the fireplace opening and must project at least 500 mm; see Figs 12.21 and 12.26.

13. The hearth must be at least 125 mm thick, formed of combustible materials.

3. The construction of a chimney through the ground floor, first floor and roof

It should be noted that solid fuel appliances are frequently omitted from upper floor levels and construction is therefore simplified (Fig. 12.22).

However, it is felt that the apprentice should be aware of traditional practice and of the problems that might be encountered in refurbishment of buildings. A back-to-back fireplace on a party wall has therefore been selected for the purpose of description, as this should cover adequately all types of fireplace construction.

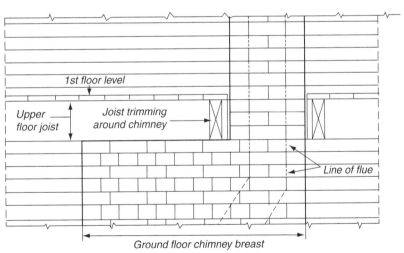

1st floor level

Upper floor joist

Joist trimming around chimney

Line of flue

Ground floor chimney breast

The above conditions only apply to timber upper floors

Figure 12.22 Construction where fireplace is not required at first floor level

Figures 12.23 and 12.24 illustrate the plan, sectional elevation and section of ground floor fireplace construction.

The chimney breast is 1.349 m in width, made up of two 330 mm jambs and a 675 mm fireplace opening, this being sufficiently wide to contain the upper floor fireplace consisting of a 225 mm jamb, a 450 mm chimney jamb containing the flue, and a 675 mm fireplace opening.

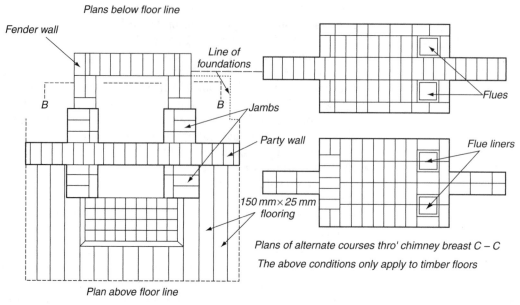

Figure 12.23 Plans of the chimney breast

Fender wall

A one-brick fender wall supports the constructional hearth and is built up on the site concrete. It is advisable to make this wall one brick in thickness, rather than a half brick, as it serves the double purpose of supporting the hearth and part of the timber floor. The void beneath the hearth should be filled with clean hardcore consisting, if possible, of broken brick; see Fig. 12.25.

The ground floor of many properties has an in-situ cast solid concrete floor or suspended flooring of pre-stressed concrete beams with $440 \times 215 \times 100$ mm blocks between. In such cases, there is no fender wall, and the whole floor is of course non-combustible, not just the minimum requirement of 500 mm in front of any fireplace.

Constructional hearth

With appliances, it is also necessary to provide a hearth in order to reduce the fire risk. Each appliance shall have a constructional hearth which shall be:

1. not less than 125 mm thick;
2. not lower than the surface of any floor built of combustible material;
3. extended within the recess to the back and jambs of the recess and projected less than 500 mm in front of the jambs and not less than 150 mm each side of the jambs;
4. not less than 840 mm square if the hearth is not constructed within a recess.

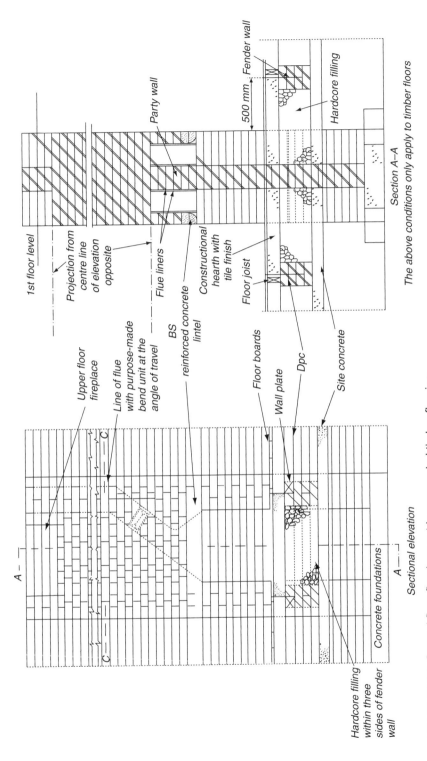

Figure 12.24 Ground floor fireplace set in suspended timber flooring

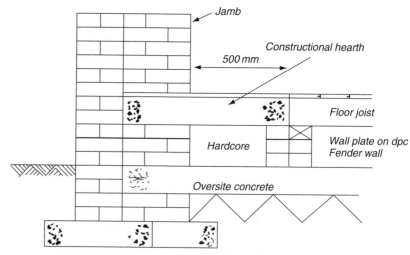

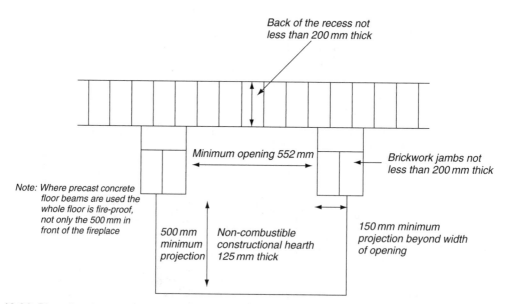

Figure 12.25 Section through a ground floor recess

The above conditions only apply to timber ground floors

Figure 12.26 Plan showing minimum dimensions for a constructional hearth

Opening for fireplace

The fire opening is a minimum of 552 mm wide. It has to be reduced at the top of the fire opening to receive a 225 mm flue liner. This can be carried out in two ways:

1. Traditional throated lintel.

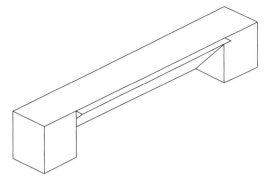

Figure 12.27 Throated lintel

The brickwork is then corbelled either side to close the opening to receive the flue liners.

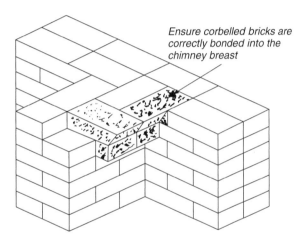

Ensure corbelled bricks are correctly bonded into the chimney breast

Figure 12.28 Corbelling to receive the flue liner

2. The lintel can be replaced by a pre-cast refractory concrete throat unit which provides access for forming throating when appliance is fitted and supports the flue liners.

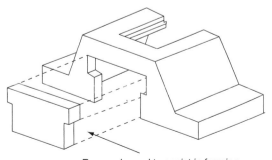

Removal panel to assist in forming the throating to flue throating unit

Figure 12.29 Throating unit

Flue liners

The path of the flue has been shown by dotted lines. Purpose-made bends of fireclay or terracotta are used to provide a continuous smooth inside surface (see Fig. 12.34), allowing a flue to be 'travelled' to left or right within the chimney breast brickwork.

Flue liners are continuously bedded and jointed ahead of the courses, and surrounding brickwork is cut and solidly bedded around them as shown in Figs 12.30 and 12.31.

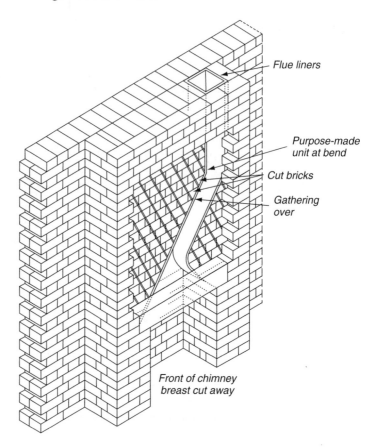

Figure 12.30 Typical chimney breast construction

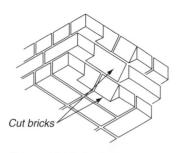

Figure 12.31 Gathering over brickwork around flue linings

Great care must be taken to see that flue liners are always set and bedded with the rebate or socket uppermost. This is to ensure that any acidic condensation which forms inside a cold flue, when a fire is lit, cannot seep out of the joints between flue liners and attack the surrounding brickwork. Figure 12.30 is an isometric detail of the gathering over. Note the arrangement of bricks to give quarter bond and a complete tying-in of the cut bricks.

The use of flue liners has overcome the problems associated with the formation of a smooth flue, using internal rendering (or parging in older buildings), before purpose-made flue liners were required by the Building Regulations after 1965.

Figure 12.32 illustrates the construction of an upper floor fireplace. It will be seen that the section has been taken through the centre of the

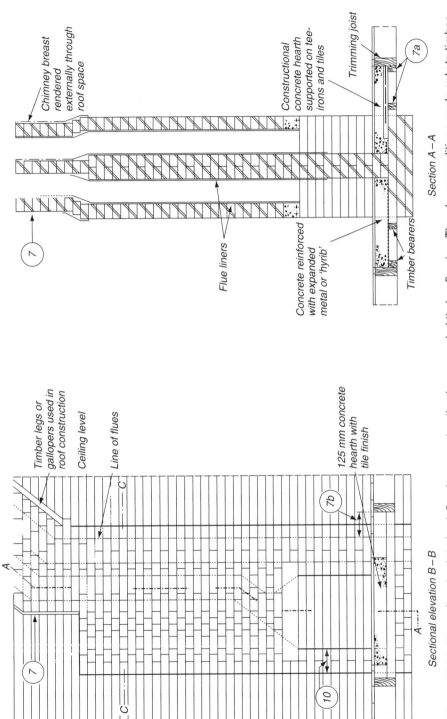

Chimney breast rendered externally through roof space

Constructional concrete hearth supported on tee-irons and tiles

Trimming joist

7a

7

Flue liners

Concrete reinforced with expanded metal or 'hyrib'

Timber bearers

Section A – A

Timber legs or gallopers used in roof construction

Ceiling level

Line of flues

A

7

125 mm concrete hearth with tile finish

7b

C

C

10

A

A

Sectional elevation B – B

Figure 12.32 Upper floor back-to-back fireplace construction in suspended timber flooring. The above conditions apply only to timber floors

fireplace and continued throughout one of the flues. This enables the construction immediately above ceiling level to be clearly illustrated, and it is also amplified in the detail (Fig. 12.33). The chimney breast has been set off to the proper stack size and can either be carried straight up or grouped with the adjacent chimney breast to form a single stack as shown in Fig. 12.35.

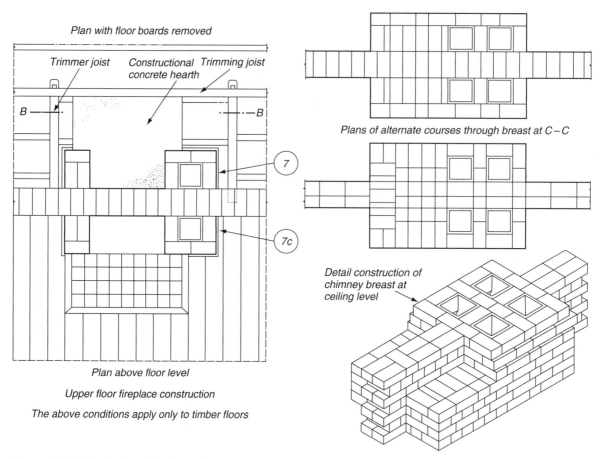

Plan with floor boards removed

Trimmer joist   Constructional Trimming joist
                concrete hearth

Plan above floor level

Upper floor fireplace construction

The above conditions apply only to timber floors

Plans of alternate courses through breast at C – C

Detail construction of chimney breast at ceiling level

Figure 12.33 Reduction of back-to-back chimney breast to chimney stack size at ceiling level

Two methods are shown for the construction of the upper floor concrete hearth; one is reinforced with expanded metal strengthened with pressed steel tees, such as 'hyrib', and the other is formed by a series of short tee-irons placed at convenient centres to receive roofing tiles (with nibs removed), or creasing tiles, which in turn support the concrete. Several methods of permanent non-combustible support are possible, but in every case the concrete must be adequately reinforced.

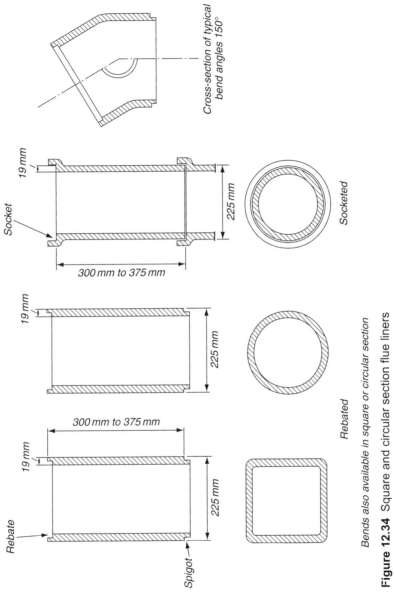

Cross-section of typical bend angles 150°

19 mm

Socket

225 mm

300 mm to 375 mm

Socketed

19 mm

225 mm

Rebated

300 mm to 375 mm

19 mm

225 mm

Rebate

Spigot

Bends also available in square or circular section

Figure 12.34 Square and circular section flue liners

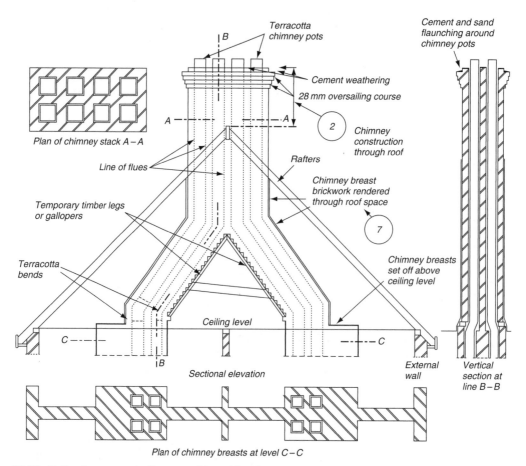

Figure 12.35 Gathering separate flues on either side of a party wall corbelling into a single chimney stack, to penetrate the roof at one point only

Figure 12.35 illustrates the chimney construction through the roof. The chimney breasts, reducing to proper size, are grouped to form a single stack. This is achieved by the use of temporary timber legs or 'gallopers', supported by a corbel projected from the chimney breast as illustrated. The section has been taken throughout one of the flues. The angled 'travel' of these flues, corbelled out from the party wall, is controlled by the purpose-made flue liner units used at the bends. Temporary timber gallopers should be erected as soon as the corbel has been sufficiently tailed down by the brickwork immediately above and has the stability to bear the weight of the gallopers.

Chimney stacks

Figure 12.36 shows the plans of alternate courses of various sized stacks, and a sectional elevation illustrating the bedding of the

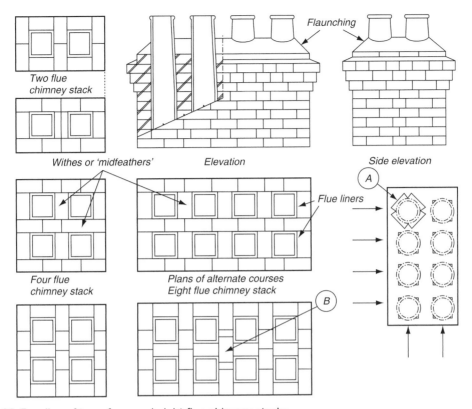

Figure 12.36 Bonding of two-, four- and eight-flue chimney stacks

chimney-pots. It is advisable to place pieces of slate directly under the pots to assist in their correct alignment and to prevent mortar from dropping down the flues where these are not completely covered by the chimney-pot (A). As the pots are usually uneven in shape, it is impracticable to line them in with a straight-edge, but they should be lined in by the eye in the direction of the arrows. This may appear superfluous, but as a line of chimney-pots can easily be seen from ground level, any lack of alignment presents an unsightly appearance and is indicative of careless workmanship.

B is an alternative method of bonding the 'withe' walls; it is used on the internal withe walls and makes use of the closers that would otherwise be wasted.

Figures 12.37 and 12.38 show two examples of finish to the top of a stack. There are many excellent examples to be seen in all parts of the country where the stacks have been carefully designed by the architect as a special feature of the building.

In Fig. 12.38 the outer walls of the stack are built in Flemish bond, one brick thick, with inner walls or 'withes' of half brick. The outer walls are more than half brick in thickness, so that excessively cold flues may to some extent be avoided.

Figure 12.37 Bonding of three-flue chimney stack

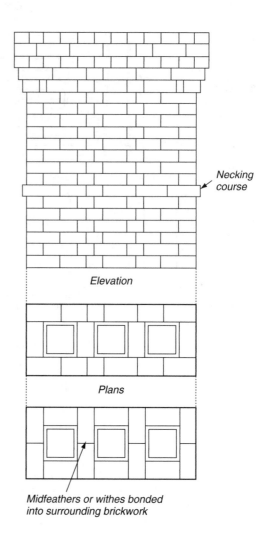

Necking course

Elevation

Plans

Midfeathers or withes bonded into surrounding brickwork

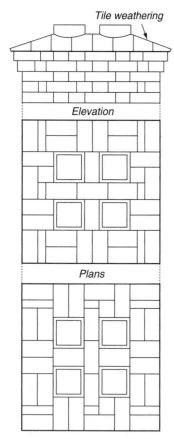

Tile weathering

Elevation

Plans

Figure 12.38 Bonding of four-flue chimney stack with one-brick thick outer walls

Figure 12.39 shows the plans of the alternate courses of bonding in an eight-flue stack, with the flues in line.

Figure 12.40 illustrates the planning of a diagonal stack; it is usual for these to be built from a rectangular base, which must be of sufficient size. It will be seen that the outer walls and withes of this rectangular base are of one brick thickness. The spaces created on the rectangular base where the diagonal stack commences are filled in with tumbled brick weatherings.

Weathering of chimney stacks passing through sloping roofs
Figure 12.41 shows a horizontal dpc of sheet lead bedded in chimney stack brickwork, and provision of lead flashings between brickwork and roof tiling.

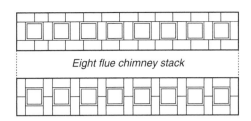

Figure 12.39 Bonding chimney stack of eight flues in line

Eight flue chimney stack

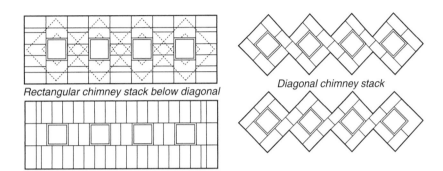

Figure 12.40 Bonding of diagonal chimney stack

Rectangular chimney stack below diagonal

Diagonal chimney stack

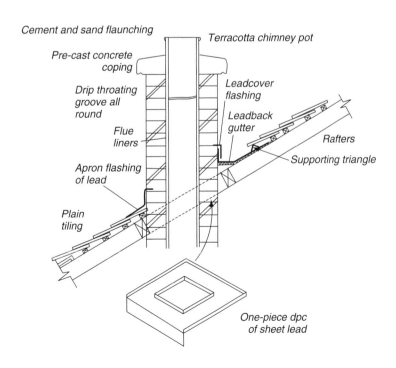

Figure 12.41 Typical construction and waterproofing at roof level

Cement and sand flaunching

Pre-cast concrete coping

Terracotta chimney pot

Drip throating groove all round

Leadcover flashing

Leadback gutter

Flue liners

Rafters

Supporting triangle

Apron flashing of lead

Plain tiling

One-piece dpc of sheet lead

4. Fireplaces

There are numerous appliances on the market which means it is impossible to cover them all in this unit. This unit will therefore attempt to cover the basic principles which apply to all appliances.

IT IS THEREFORE IMPORTANT TO READ THE MANUFACTURER'S INSTRUCTIONS AT ALL TIMES.

Heating appliances may be classified into the following groups:

- Open fires
- Inset fires
- Inset with underfloor primary air supply
- Open fires with back boilers
- Convector open fires
- Free-standing convector open fire
- Independent boilers.

This section will deal with open fires only.

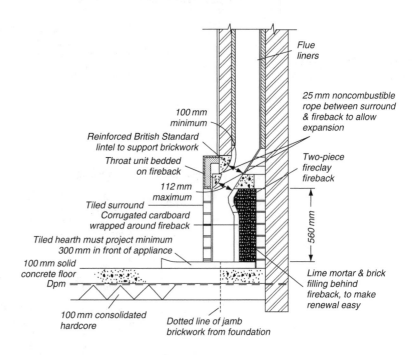

Figure 12.42 Section through a ground floor fireplace

The main components include the fire surround and tiled hearth, the fireback throat unit and fire grate (Fig. 12.43).

The fire surround and hearth should be fixed to a prepared chimney breast. The opening should be 338 mm deep and 572 mm wide. The first operation is usually to check all the sizes and ensure you have all the components and materials ready, and then to fix the fireback.

Firebacks are available in two types.

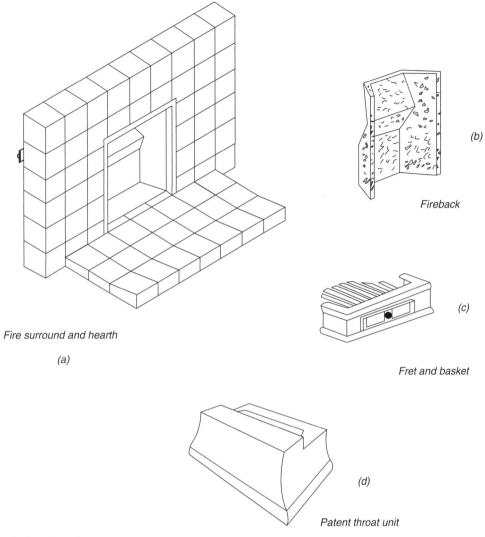

Fire surround and hearth

(a)

Fireback

(b)

Fret and basket

(c)

Patent throat unit

(d)

Figure 12.43 Identification of main components

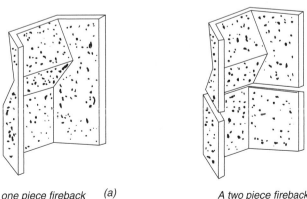

Figure 12.44 Types of fireback   *A one piece fireback*   *(a)*   *A two piece fireback*   *(b)*

Stage 1

The back hearth has to be raised to the thickness of the superimposed hearth to allow the fireback to sit at hearth level. Fire bricks should be used bedded in a lime mortar (Fig. 12.45).

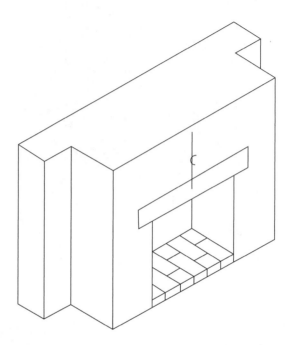

Figure 12.45 Stage 1 Fixing the back hearth

Stage 2

Place the fireback onto the prepared brickwork, centre it and check for plumb and level. You need to place the fire surround in position to check the position of the fireback. The fireback may need to project up to the surround but leaving a small gap for the expansion joint (Fig. 12.46).

Stage 3

Before building in around the fireback corrugated paper should be positioned around the rear of the back to allow for expansion. When the paper has burnt with the heat of the fire a small gap will be formed which will allow the fireback to expand when hot.

Build up the gap between the fireback and the chimney breast with bricks and weak mortar.

The void at the back should be filled with vermiculate concrete or brick rubble: 1 part cement, 2 parts lime, 8 parts broken brick. Complete this operation until the top of the fireback is reached (Fig. 12.47).

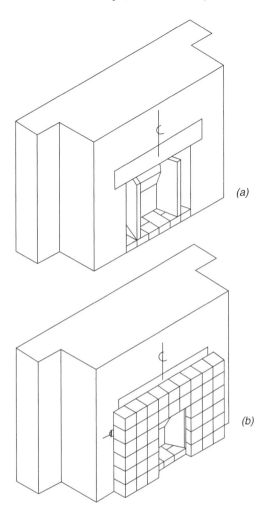

(a)

(b)

Figure 12.46 Stage 2 Fixing the fireback

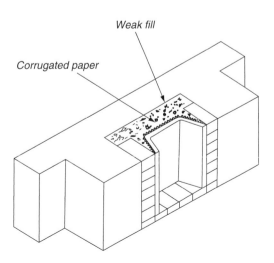

Weak fill

Corrugated paper

Figure 12.47 Stage 3 Building in around the fireback

Stage 4

The space from the top of the fireback to the flue liner must be finished with a smooth surface about 45°.

The gap or throat should be about 100 mm across to give maximum efficiency to the fire and not restrict the flue gases (Fig. 12.48).

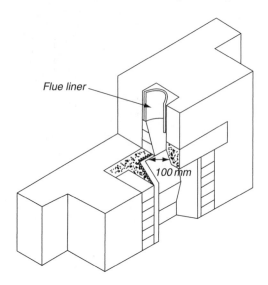

Figure 12.48 Stage 4 Forming the throat

Stage 5

Lift the surround into position and plumb and level. Make sure the expansion joint is lined up between the fireback and surround (Fig. 12.49). Mark out the position of the lugs and drill the chimney breast using a hammer drill. Screw with brass screws to secure surround.

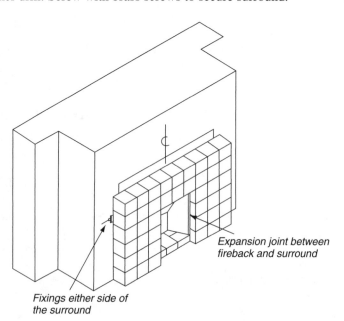

Figure 12.49 Stage 5 Fixing the tile surround

Stage 6

Bed the hearth in position using a weak lime/cement mortar. Check for level along front and width. Ensure the expansion joint is in position between the back of the hearth and the surround. Complete the joint between the hearth and surround. Wipe down all parts of the surround and hearth (Fig. 12.50).

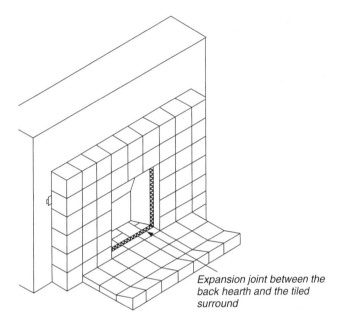

Expansion joint between the back hearth and the tiled surround

Figure 12.50 Stage 6 Fixing the tiled hearth

13 External works

External works brickwork

Garden or boundary walling, paving and steps, retaining walls or ramps and planter boxes are all examples of external works brickwork. This use of brickwork is sometimes referred to as 'hard landscaping'.

Bricks and mortar in any of these situations are very much more at risk from frost damage than the cavity walling of a house.

The outer leaf brickwork in Fig. 13.1 is nicely sheltered under a pitched roof. Any wetting from wind-blown rain is on to the outer

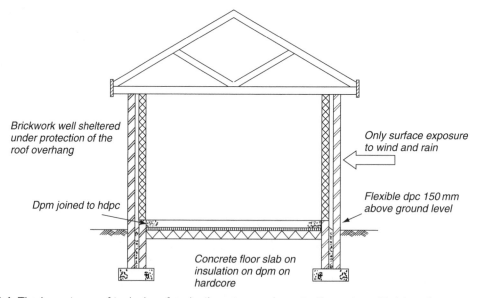

Brickwork well sheltered under protection of the roof overhang

Only surface exposure to wind and rain

Dpm joined to hdpc

Flexible dpc 150 mm above ground level

Concrete floor slab on insulation on dpm on hardcore

Figure 13.1 The importance of typical roof projection at eaves in protecting external brickwork

surface only and can quickly dry out so that bricks are unaffected by frost in winter.

The garden wall shown in Fig. 13.2 is totally exposed to rain on both sides and the top and therefore much more at risk from the effects of frost when bricks are saturated.

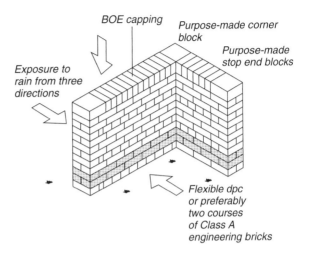

Figure 13.2 Free-standing wall with flush brick on edge capping

Selection of bricks for external works

Bricks that soak up a lot of water are more likely to suffer from the effects of frost when in a saturated state than bricks that absorb very little water. Class 'A' engineering quality bricks, due to their method of manufacture, absorb no more water than 4.5% by weight, and so are unaffected when exposed to freezing conditions. (See Chapter 2 for definitions of engineering bricks.) Some facing bricks absorb as much as 20% of their weight in water, and if frozen in this condition, will begin to crumble away at the surface.

Great care must be taken with the choice of brick for external works. Engineering quality or other 'frost proof' bricks must be specified if brickwork is to have long-term durability. From an appearance point of view well-burned stock bricks blend well with gardens and landscaping. If 'clamp burnt stocks' are specified, only first- or second-hard selected bricks should be used for landscape brickwork constructions. Always check with the manufacturer before using any brick for external works.

Mortar

The choice of mortar for external works or garden brickwork must be as carefully considered as the bricks, if it also is to be durable.

Group (ii) mortar $1:\frac{1}{2}:4\frac{1}{2}$ will give increased resistance to the effects of frost, if used with hard burnt stocks and Class 'B' engineering bricks

Group (i) mortar $1:\frac{1}{4}:3$ complements Class 'A' engineering bricks, with its higher compressive strength when hardened, and therefore has greater resistance to water penetration (see Chapter 2 for mortar designations).

Brick paving

Paving bricks or 'pavers' are classed as frost proof because they are likely to be damp throughout the year, and are particularly exposed to frost damage in winter.

Method of laying pavers

There are two ways of laying paving bricks:

(i) solid bedding on to a previously laid concrete base slab using bricklaying mortar below and between; see Fig. 13.3.

(ii) flexible bedding. This is the most commonly used and most economical way of laying paving bricks, as it dispenses with the need for a concrete foundation slab and neither is bedding mortar required; see Fig. 13.4.

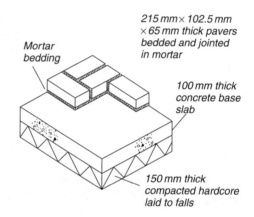

Figure 13.3 Brick paving: solid bedding method

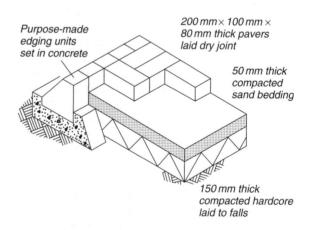

Figure 13.4 Brick paving: flexible bedding method

When laying bricks by the flexible bed method, pavers are placed upon a carefully levelled layer of sand 50 mm thick, and settled in with a mechanical plate compactor. Dry sand is brushed over areas of completed brick paving to fill any gaps. Rainwater will drain down between pavers initially with this method of laying and soak away through the hardcore sub-base.

Pathways, patios and larger areas of brick paving must always be 'laid to falls', whichever bedding method is specified, in order to prevent puddles forming in wet weather. Although initially, rainwater drains away through the 'dry joints' of flexible bedding, these soon become sealed up with dust and dirt and rainwater must 'run off' paving to falls or gradients which should be not less than 25 mm per 3 m; see Fig. 13.5.

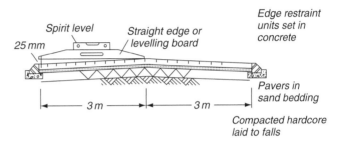

Figure 13.5 The importance of typical roof projection at eaves in protecting external brickwork

Various patterns of brick paving are possible as shown in Fig. 13.6, some using 200 mm × 100 mm size paving units, others using 215 mm × 102.5 mm purpose-made pavers, or frost proof bricks laid 'frog down' or on-edge.

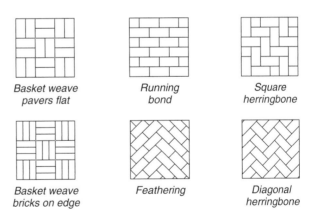

Figure 13.6 Typical patterns for brick paving

Boundary walls

These are described as 'free standing' because the brickwork derives no support from floors or roof structure as that used in a building.

For this reason Local Authority Building Control Departments' approval is usually required if a wall higher than 1 metre is to be built next to a public road or 2 metres high elsewhere, in order to check that the design is stable.

Corners, attached piers and 'breaks' in a free-standing wall provide support to resist overturning and failure at the weakest point, where a flexible dpc has been used.

Attached piers must be properly bonded to the wall as shown in Fig. 8.13.

Damp proof coursing for boundary walls is best provided by using two courses of black or red Class 'A' engineering bricks, bedded and jointed in cement mortar.

The brickwork above and below is solidly bonded to these dpc bricks, as shown in Fig. 13.8. The result bonds better than if separated by a flexible dpc, which in a free-standing boundary wall creates a plane of weakness.

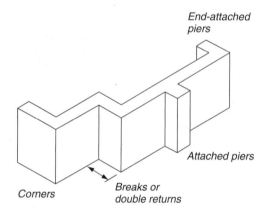

End-attached piers

Attached piers

Corners

Breaks or double returns

Figure 13.7 Ways of improving stability of free-standing walls

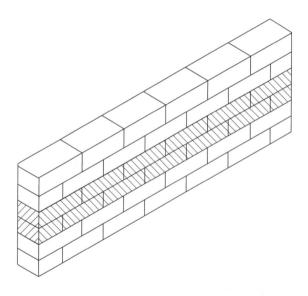

Figure 13.8 Engineering bricks used as dpc

Serpentine walling

Serpentine walling of either 102.5 mm stretcher bonded brickwork or 225 mm English or Flemish bonded brickwork is a very effective shape to resist overturning, and is normally found in boundary walls. See Chapter 8, Fig. 8.76 and Fig. 13.9.

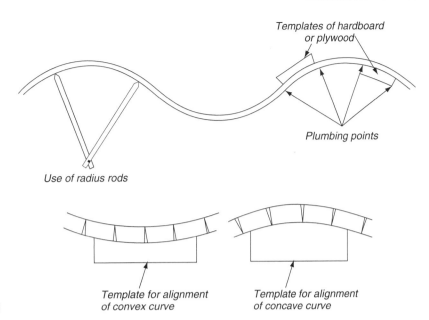

Figure 13.9 Serpentine walling

Vertical movement joints

The bricklayer must make allowances, when constructing boundary walls, for the expansion and contraction of the brickwork due to changes in air temperature and/or moisture content of the bricks.

A completely straight joint should be formed throughout the full height and thickness of the brick wall. These vertical movement joints should be 10 to 15 mm wide on the face, and are needed at approximately 9-metre intervals to prevent stress from developing in the wall due to expansion. A soft filler of expanded plastic foam strip is built in, to stop mortar bridging the 10 to 15 mm space.

To reinforce the obvious weakness caused by these unbonded, vertical joints, special stainless steel 'slip-ties' are built in every fourth course as the work proceeds, as shown in Fig. 13.10, to keep the ends of the walling permanently in line. Standard cross-cavity wall ties must not be used at these points.

The expanded plastic foam filler is kept back (or cut back), 12 mm from both faces of the brickwork, so that a sealant mastic can be applied, to seal these movement joints against rain penetration.

It is very important that vertical movement joints are taken right up through any BOE or pre-cast coping or capping on top of the wall. Generally speaking, vertical movement joints should commence at ground level, because the temperature below is practically constant.

Note: Compressed fibre board is unsuitable as a vertical movement joint filler in brick walling. It is too hard, and does not compress easily when the walling expands.

Construction

It is very important that vertical movement joints are truly plumb, straight and parallel 10 to 15 mm width on face. A good way of ensuring these

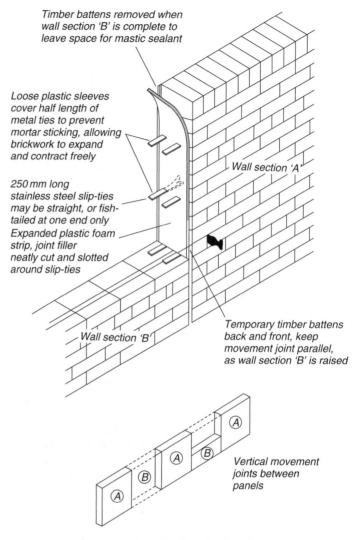

Timber battens removed when wall section 'B' is complete to leave space for mastic sealant

Loose plastic sleeves cover half length of metal ties to prevent mortar sticking, allowing brickwork to expand and contract freely

250 mm long stainless steel slip-ties may be straight, or fish-tailed at one end only
Expanded plastic foam strip, joint filler neatly cut and slotted around slip-ties

Wall section 'A'

Wall section 'B'

Temporary timber battens back and front, keep movement joint parallel, as wall section 'B' is raised

Vertical movement joints between panels

Figure 13.10 Formation of vertical movement joint

Sequence of construction of wall sections

three important requirements is to build the wall in alternate sections between vertical movement joint locations, as shown in Fig. 13.10.

The stopped ends of wall sections (A) can be accurately plumbed and built to gauge first. When wall sections (B) are in-filled, courses being run-in to the line, using 10 to 15 mm thick temporary timber slips to leave truly straight and parallel vertical joints upon completion of the wall.

Copings and cappings

The top of free-standing walls must be protected as effectively as is possible, to prevent downwards penetration of rainwater. A continuous top covering which looks attractive and acts like a good umbrella will give the best long-term protection to a free-standing wall and avoid damage to brickwork by frost.

Figure 13.11 shows the desirable features of a well-designed coping. Flush cappings lack some of these important features and so afford less effective protection; see Figs 13.2. and 13.13.

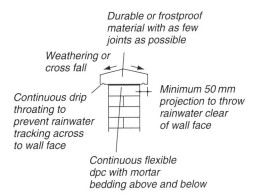

Durable or frostproof
material with as few
joints as possible

Weathering or
cross fall

Continuous drip
throating to
prevent rainwater
tracking across
to wall face

Minimum 50 mm
projection to throw
rainwater clear
of wall face

Continuous flexible
dpc with mortar
bedding above and below

Figure 13.11 Requirements of an effective durable coping

Continuous lengths of non-ferrous metal or plastic would provide the best sort of protection from rain, due to having very few joints, but these materials are not commonly used.

Lengths of pre-cast concrete, typically 750 mm long and saddleback or feather edge in cross section embody the best features of the well-designed coping as indicated in Fig. 13.11, but leakage can occur at the mortar joints.

Natural stone copings of similar cross section are much more expensive as these are sawn from naturally quarried Portland or other durable stone.

Plain brick-on-edge copings and cappings, although traditional and very commonly used, present many mortar joints through which rain can penetrate; see Fig. 11.17.

Figure 13.12 shows lime staining on a boundary wall where lime is leaching from mortar under wet conditions. Unsightly patches of lime are left underneath each joint in the concrete copings.

All stone, pre-cast concrete and brick copings and cappings should be bedded upon a continuous dpc to arrest any leaking at joints; see Fig. 13.11. Various patented extensions of the basic brick-on-edge are available, which seek to make copings and cappings more secure; see Fig. 13.14.

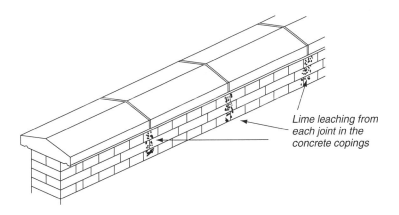

Lime leaching from
each joint in the
concrete copings

Figure 13.12 Lime leaching

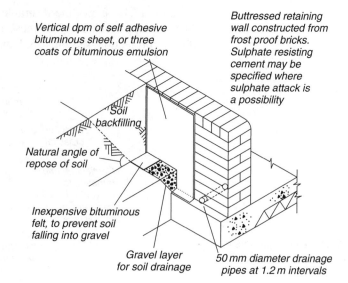

Vertical dpm of self adhesive bituminous sheet, or three coats of bituminous emulsion

Buttressed retaining wall constructed from frost proof bricks. Sulphate resisting cement may be specified where sulphate attack is a possibility

Soil backfilling

Natural angle of repose of soil

Inexpensive bituminous felt, to prevent soil falling into gravel

Gravel layer for soil drainage

50 mm diameter drainage pipes at 1.2 m intervals

Figure 13.13 Damp proofing and drainage behind a brick retaining wall

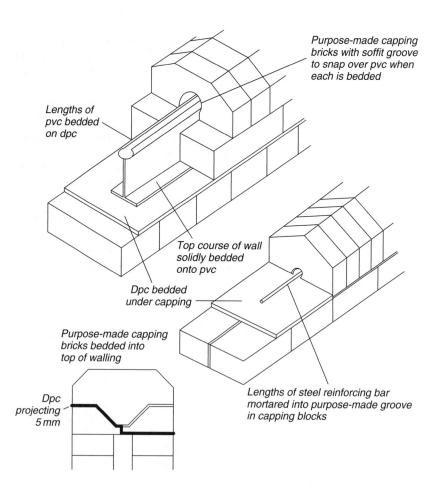

Purpose-made capping bricks with soffit groove to snap over pvc when each is bedded

Lengths of pvc bedded on dpc

Top course of wall solidly bedded onto pvc

Dpc bedded under capping

Purpose-made capping bricks bedded into top of walling

Dpc projecting 5 mm

Lengths of steel reinforcing bar mortared into purpose-made groove in capping blocks

Figure 13.14 Ways of making a brick-on-edge capping course more secure

Brick retaining walls

Any walling intended to hold back soil must be properly designed by a structural engineer, so as to resist overturning from sideways pressure. Even a 215 mm thick wall that is only six courses high and intended to hold back a soil bank may begin to topple after a couple of years; see Fig. 13.15.

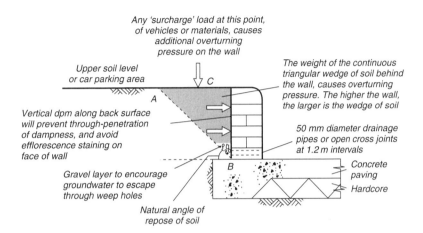

Figure 13.15 Cross section through low retaining wall, showing typical sideways pressure from soil behind

The line AB in Fig. 13.15 represents a typical natural angle of repose for soil, say 45 degrees. Motorway embankments are frequently seen left at this natural slope with the soil completely stable with grass and bushes growing on them. The angle of repose, of course, depends on the type and condition of the soil.

It is the 'soil wedge' above line AB in Fig. 13.15 that presses against the vertical back of a retaining wall, along its full length, which is tending to push over the brickwork. Retaining walls must be able to permanently resist the pressure of this triangular mass of soil and remain plumb.

Any extra load at point C in Fig. 13.15, i.e. stacks of materials or vehicles, is called a surcharge, and increases the overturning pressures on the retaining wall.

Retaining wall construction

Figure 13.16 shows a number of ways that a brick retaining wall can be strengthened against sideways or lateral pressure.

Class 'A' or 'B' engineering quality or other frost proof bricks should be specified, using a Group (i) or Group (ii) mortar, to ensure long-term durability.

The soil side of a brick retaining wall should be covered with a dpm (damp proof membrane) to prevent through-penetration of groundwater, which can cause unsightly efflorescence or other salt staining of the outer surface. This dpm can be formed from 900 mm wide rolls of paper-backed

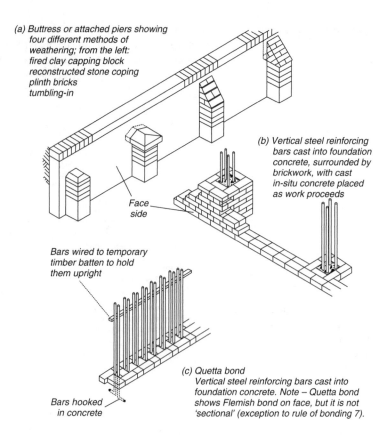

(a) Buttress or attached piers showing four different methods of weathering; from the left:
fired clay capping block
reconstructed stone coping
plinth bricks
tumbling-in

Face side

(b) Vertical steel reinforcing bars cast into foundation concrete, surrounded by brickwork, with cast in-situ concrete placed as work proceeds

Bars wired to temporary timber batten to hold them upright

Bars hooked in concrete

(c) Quetta bond
Vertical steel reinforcing bars cast into foundation concrete. Note – Quetta bond shows Flemish bond on face, but it is not 'sectional' (exception to rule of bonding 7).

Figure 13.16 Three ways that a structural engineer may consider strengthening a brick retaining wall

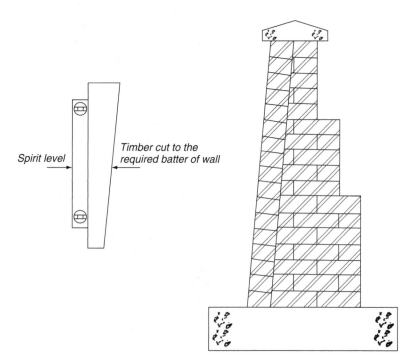

Spirit level

Timber cut to the required batter of wall

Figure 13.17 Section through retaining wall with battered face to front and stepped at rear

self-adhesive bitumen, pressed firmly on to the brickwork at the soil side of the wall.

Alternatively, three separate coats of bituminous emulsion paint may be applied to the back surface of the brickwork to form a vertical dpm; see Fig. 13.13.

Battering walls

Retaining walls can be designed with battering faces. A special battering plumb rule is required to erect the battering face of the wall. Alternatively, certain types of spirit levels have adjustable vertical spirit tubes which can be used.

Buttresses

When a wall requires additional lateral support a buttress can be designed and constructed at intervals along its face to provide more stability as shown in Fig. 13.18. Lateral forces can cause a wall to buckle so buttresses can be attached to the wall at intervals along its length to increase stability. These buttresses will generally have to resist greater stresses at the base of the wall than at the top of the wall. This is the reason for the design of the buttresses which are wider at the bottom than at the top. The buttress should be properly bonded to the main wall.

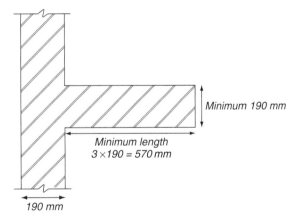

Figure 13.18 Building Regulation requirements for buttressing walls

The buttress can be built by battering the face or a stepped face, reducing the width with the use of plinth bricks or tumbling-in.

A buttress should be built from the base of the main wall to a distance from the top of the main wall equal to three times the thickness of the main wall it is supporting (including the thickness of the main wall itself). The projection of the buttress should be equal to at least three times the thickness of the main wall and should be at least 190 mm wide.

Drainage

Small drainpipes should be built into the lowest course of a brick retaining wall to prevent any build up of static water pressure. A layer of hard-core or gravel behind the retaining wall should be spread between these drainage pipes, before the soil is backfilled, upon completion of the wall; see Fig. 13.13.

Inspection chambers

Brickwork in inspection chambers is below ground and out of sight but it must still be built correctly as the inspection chamber will have to be tested against water leaking out.

This section only provides a basic guide and operations could change around the country.

Inspection chambers are required for access to the drainage and sewage systems. Brick chambers could also be required for access to other services.

The basic requirements of an inspection chamber are that it should be large enough to allow a person entrance and to work. There could be branch drains entering the inspection chamber so it should be designed according to the number of branches. It has to be watertight and resist all possible loads from above and externally.

The recommended minimum internal sizes are:

450 mm long × 450 mm wide when the inspection chamber is not more than 1 metre deep; see Fig. 13.20.

1200 mm long × 750 mm wide if more than 1 metre deep; see Fig. 13.21.

When inspection chambers have branches there should be a minimum of 300 mm allowed for each 100 mm and 150 mm branch on the side having the most branches. It is also advisable to allow 600 mm at the outlet end of the inspection chamber to allow for further connections to the inspection chamber and for easy rodding. The minimum width should be 450 mm or an extra 300 mm if there is a branch.

The inspection chamber should be constructed on a concrete base, normally 150 mm thick and built with one brick thick walls in cement mortar. Some authorities may specify Class B clay engineering bricks and sulphate resisting cement if the ground contains sulphates.

The inspection chamber is normally built in English bond but if there is high water pressure from outside, 'water bond' may be specified; see Fig. 13.24.

If the inspection chamber is over 750 mm deep step irons should be built in every fourth course beneath the opening. They should be staggered, 300 mm apart, to allow easy access. Figure 13.22 shows a typical galvanised step iron.

The cover frame should be either 450 mm × 450 mm for inspection chambers less than 1 metre deep or 600 mm × 600 mm for those more than 1 metre deep.

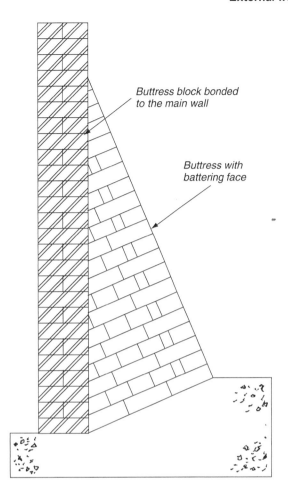

Figure 13.19 Section through a buttress with a battering face

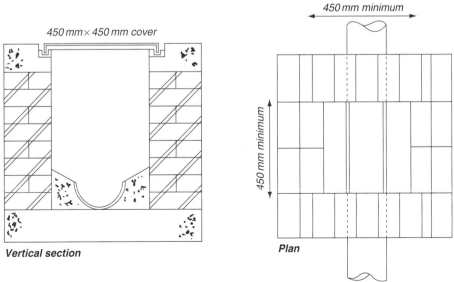

Figure 13.20 Inspection chamber less than 1 metre deep

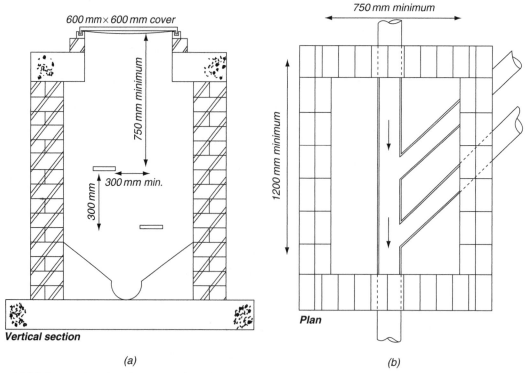

600 mm × 600 mm cover

750 mm minimum

300 mm

300 mm min.

Vertical section

(a)

750 mm minimum

1200 mm minimum

Plan

(b)

Figure 13.21 Inspection chamber more than 1 metre deep

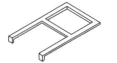

Figure 13.22 Galvanised step iron

Construction of inspection chambers

The main channel pipe is laid when the main run of pipes is being laid by the drain layers. Any connecting branches are also laid and completed. The concrete base is then laid in position around the channel pipes and allowed to set. The brickwork is set out for the inspection chamber according to the number of branches required. Approximately six courses should be built taking care to keep the inspection chamber plumb, level and square. Any pipes through the walls should be protected from any pressure by turning small arches over those of 150 mm and above; see Fig. 13.23. The benching should be completed at this level when it is easier.

Bricks could be used to form the basic structure of the benching and then finished smoothly with a stiff concrete. The gradient should be approximately 1:12 to prevent any waste from congregating on the benching; see Fig. 13.25. Once the benching has hardened the remainder of the brickwork can proceed. If the inspection chamber is over 750 mm deep step irons should be built in to assist in access to the chamber.

The inspection chamber is usually roofed over with an in-situ reinforced concrete slab with an access hole.

The brick shaft can be extended until the correct height is reached and the inspection cover fitted.

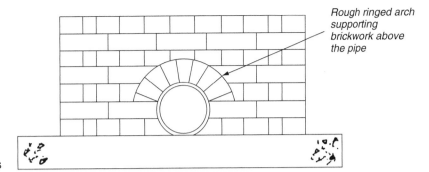

Figure 13.23 Arches over pipes

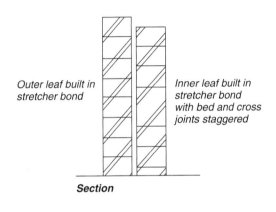

Figure 13.24 Water bond details

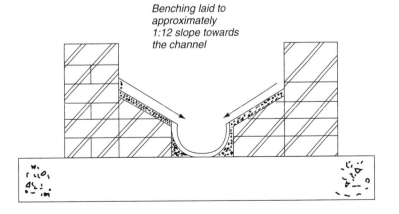

Figure 13.25 Benching details

Smaller inspection chambers could have the brickwork corbelled inwards to receive the inspection cover instead of a concrete slab.

Brick inspection chambers have been superseded on many construction sites by concrete pre-cast concrete inspection chambers which are circular on plan. The bricklayer may still be required to construct the small chamber on top of the concrete inspection chamber to receive the inspection cover.

14 Special shaped bricks

The bulk of production by brick manufacturers is of plain, basic metric bricks, with a work size of $215 \times 102.5 \times 65$ mm. These are produced as solid, frogged or perforated, from clay, sand lime or concrete as described in Chapter 2 of this book.

Manufacturers do, however, make a wide range of other brick shapes, and some examples are shown in Fig. 14.1.

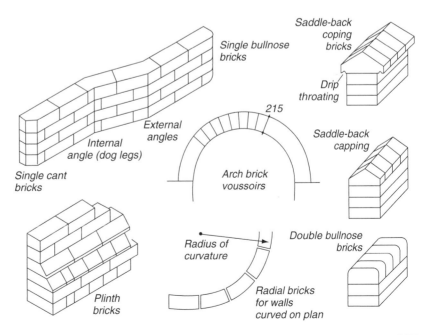

Figure 14.1 Uses for special bricks

BS 4729:1990 describes these variants as 'Bricks of Special Shapes and Sizes'. The majority are based upon the dimensions of the basic metric brick (see Fig. 14.2), so they will fit in with normal bonding arrangements and vertical gauge. These are given a type number and prefix number in BS 4729, to make ordering a simple matter.

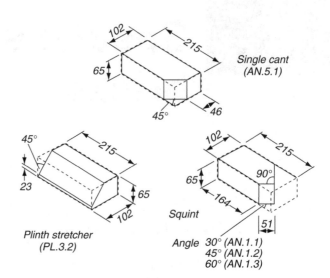

Figure 14.2 Enclosing dimensions of typical special shape bricks

Bricklayers know the most commonly used ones as 'standard specials', intended for the decorative and functional purposes required of brickwork – see Fig. 14.3. They usually refer to all other shapes that might be specified by an architect as 'special specials'. These require detailed drawings to be sent to the brick manufacturer, so that a mould can be made.

Availability of special shaped bricks

Although a brick of special shape might be referred to as a 'standard special', it will not necessarily be held in stock by the brick manufacturer or builders' merchant. Since the hundreds of different colours and surface textures of facings each have their own respective range of special shapes, stocking them all would be a very expensive business.

The BS 4729:1990 divides special bricks into ten groups for convenience of classification. Each group has recognisable prefix letters to make the system user-friendly, see Fig. 14.4, which shows a selection from the total range available.

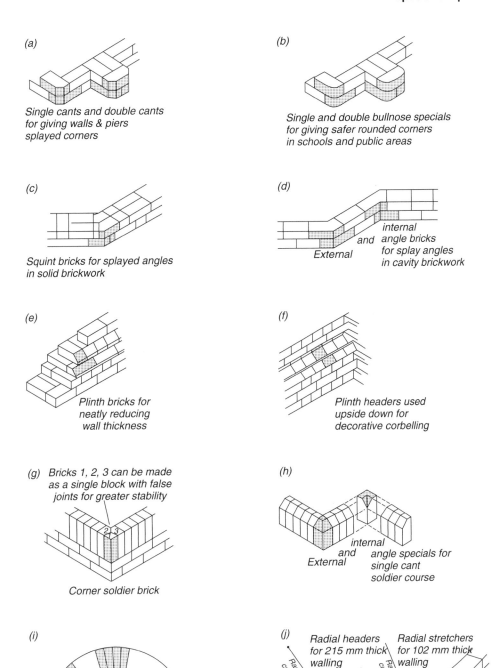

(a)

Single cants and double cants
for giving walls & piers
splayed corners

(b)

Single and double bullnose specials
for giving safer rounded corners
in schools and public areas

(c)

Squint bricks for splayed angles
in solid brickwork

(d)

internal
and *angle bricks*
for splay angles
External *in cavity brickwork*

(e)

Plinth bricks for
neatly reducing
wall thickness

(f)

Plinth headers used
upside down for
decorative corbelling

(g) Bricks 1, 2, 3 can be made
as a single block with false
joints for greater stability

Corner soldier brick

(h)

internal
and *angle specials for*
External *single cant*
soldier course

(i)

Arch brick
voussoirs

(j)

Radial headers   Radial stretchers
for 215 mm thick   for 102 mm thick
walling   walling

Radius of curvature   *Radius of curvature*

Built   Built
header bond   stretcher bond

Figure 14.3 Application of special shape bricks

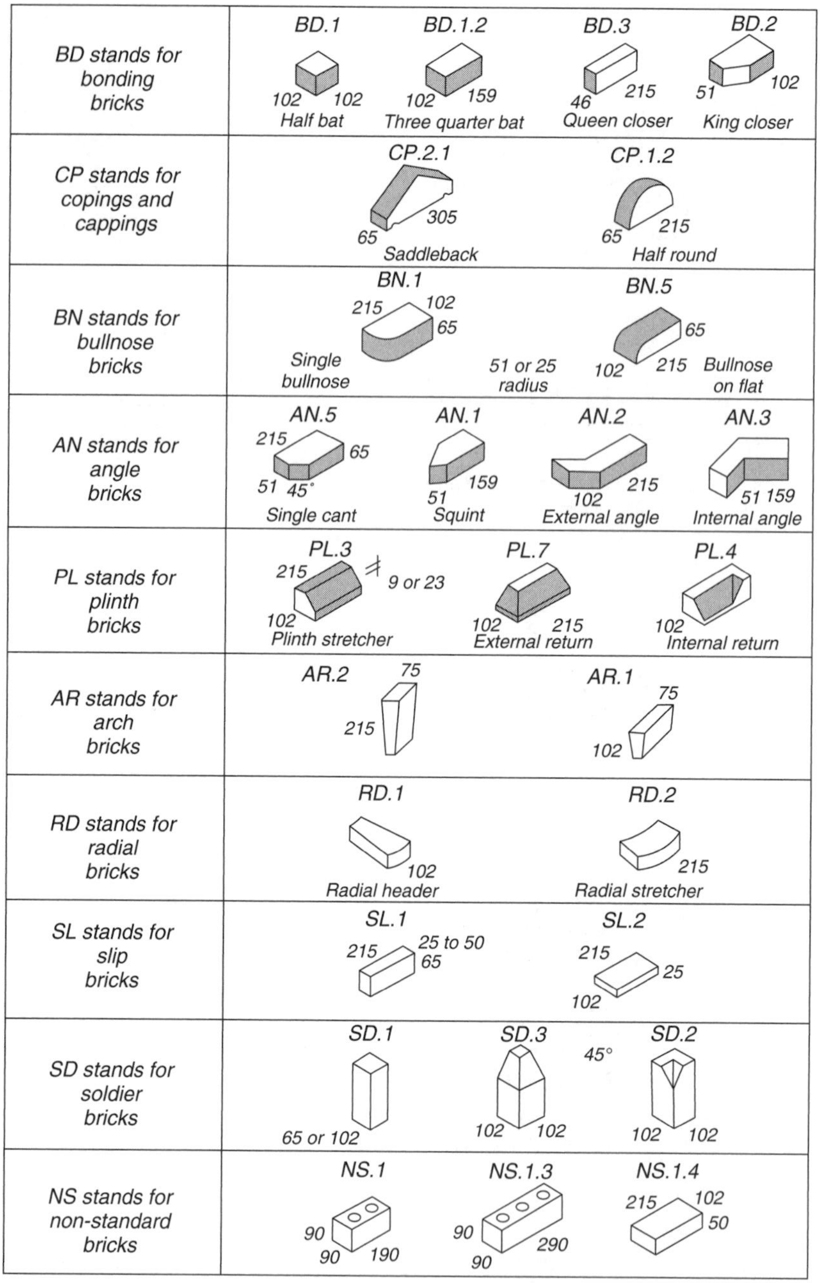

BD stands for bonding bricks	BD.1 — 102 102 — Half bat	BD.1.2 — 102 159 — Three quarter bat	BD.3 — 46 215 — Queen closer	BD.2 — 51 102 — King closer
CP stands for copings and cappings		CP.2.1 — 65 305 — Saddleback	CP.1.2 — 65 215 — Half round	
BN stands for bullnose bricks		BN.1 — 215 102 65 — Single bullnose — 51 or 25 radius	BN.5 — 65 102 215 — Bullnose on flat	
AN stands for angle bricks	AN.5 — 215 65 51 45° — Single cant	AN.1 — 51 159 — Squint	AN.2 — 102 215 — External angle	AN.3 — 51 159 — Internal angle
PL stands for plinth bricks	PL.3 — 215 102 9 or 23 — Plinth stretcher	PL.7 — 102 215 — External return	PL.4 — 102 — Internal return	
AR stands for arch bricks	AR.2 — 75 215 —	AR.1 — 75 102 —		
RD stands for radial bricks	RD.1 — 102 — Radial header	RD.2 — 215 — Radial stretcher		
SL stands for slip bricks	SL.1 — 215 25 to 50 65 —	SL.2 — 215 25 102 —		
SD stands for soldier bricks	SD.1 — 65 or 102 —	SD.3 — 102 102 —	SD.2 — 45° 102 102 —	
NS stands for non-standard bricks	NS.1 — 90 90 190 —	NS.1.3 — 90 90 290 —	NS.1.4 — 215 102 50 —	

Figure 14.4 BS 4729:1990 Classification (*Note*: dimensions shown in mm are net brick sizes, without joint allowances)

Stop bricks

Certain special bricks are shaped so that, for example, a bullnose effect can be changed neatly back to a square corner. Examples of different kinds of 'stop bricks' are shown in use in Figs 14.5 and 14.6.

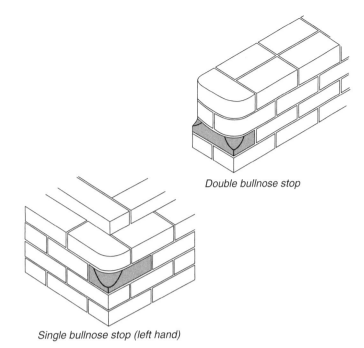

Double bullnose stop

Single bullnose stop (left hand)

Figure 14.5 Uses of special shape bullnose stop bricks

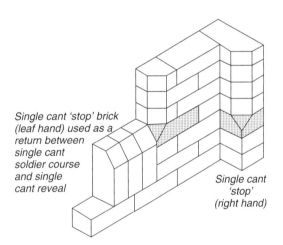

Single cant 'stop' brick (leaf hand) used as a return between single cant soldier course and single cant reveal

Single cant 'stop' (right hand)

Figure 14.6 Uses of special shape cant stop bricks

Bonding with obtuse angle specials

Squint bricks (AN.1)

Traditionally, squint bricks are produced specifically for bonding obtuse angle quoins in *solid* brick walling (see Fig. 14.7). The dimensions of each face of the squint brick shown, allow $1/4$ lap to be maintained on face. However, the bricklayer may be called upon to use squint bricks in 102 mm thick walling, in some modern building – although this is not good practice because the $1/2$ lap of stretcher bonding is not being maintained around the obtuse angle quoin.

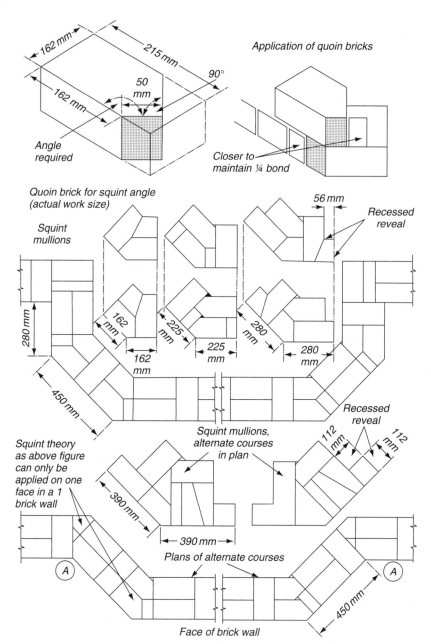

Figure 14.7 Bonding squints

External angle (AN.2) and internal angle (AN.3) specials

These special shaped bricks, illustrated in Fig. 14.3(d), are intended for bonding obtuse angle quoins in stretcher bond (cavity) walling. The face dimensions of each allow $1/2$ lap to be continued around splayed angles.

Note: Internal angle specials (AN.3), sometimes referred to as 'dog legs', are faced on the sides opposite to external angle special bricks (AN.2).

Arch bricks

A trained bricklayer is able to set out an axed arch (see page 187), and cut the required number of voussoirs from basic-size bricks using hammer, bolster and comb hammer.

BS 4729:1990 includes four standard shape arch bricks, however, which manufacturers will mould to a tapering shape, suitable for set spans for semicircular arches of 910, 1360, 1810 and 2710 mm, see Fig. 14.8. Arches of any other span or shape will require special shape arch bricks, to obtain parallel mortar joints between voussoirs, and would need to be specially ordered.

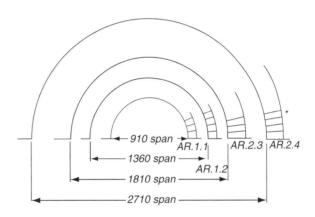

Figure 14.8 Arch bricks. Elevation showing four standard spans all available in BS 4729 as either tapered headers or tapered stretchers

Extrados dimension with all four types 75 mm

Brick on edge quoin blocks, angles and stopped ends

To avoid cutting mitres at the point where a BOE capping turns a corner or obtuse angle, a range of fired clay 'blocks' are made for the purpose. Figure 14.9 shows that these BS 4729 specials are made to suit bullnose and cant as well as plain BOE finish. All are very secure when firmly bedded, providing solid support for intermediate bricks.

'Handing' of special shaped bricks

If the single cant special bricks, shown in Fig. 14.3(a) have a smooth face and are solid wirecuts, then they can be used as the 'quoin bricks' in *every* course simply by being turned over in alternate courses. They do not have a top and a bottom as such.

Depending upon the method of manufacture, some specials will have a single frog, which should always be laid 'frog-up'. Other facing bricks

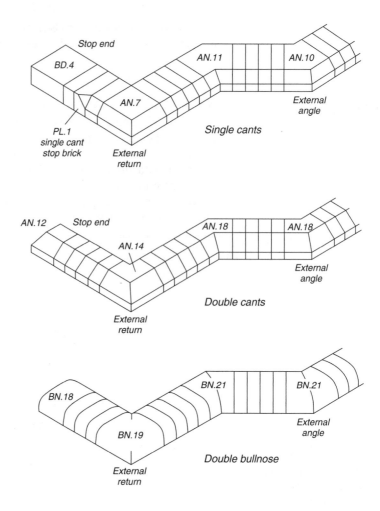

Figure 14.9 BS 4729:1990
Brick cappings

may be perforated wirecuts with a 'dragwire' surface texture, and these must *not* be turned over either. This is because the surface texture is directional, and the bricks must *not* be bedded so that rain is retained at the surface.

Therefore, special shape bricks, like those shown in Fig. 14.3(a) to (d) inclusive, have a definite top and bottom, due to the presence of a single frog or a dragwire surface texture. 'Left handed' and 'right handed' versions must be ordered, as indicated in Fig. 14.10.

Similarly, can you imagine trying to use the single bullnose and single cant stop bricks illustrated in Figs 14.5 and 14.6 respectively if the change back to a square angle was required on the course *below* that shown in the illustrations? For this reason, stop bricks must also be ordered left and right handed to suit requirements, as indicated in Fig. 14.11.

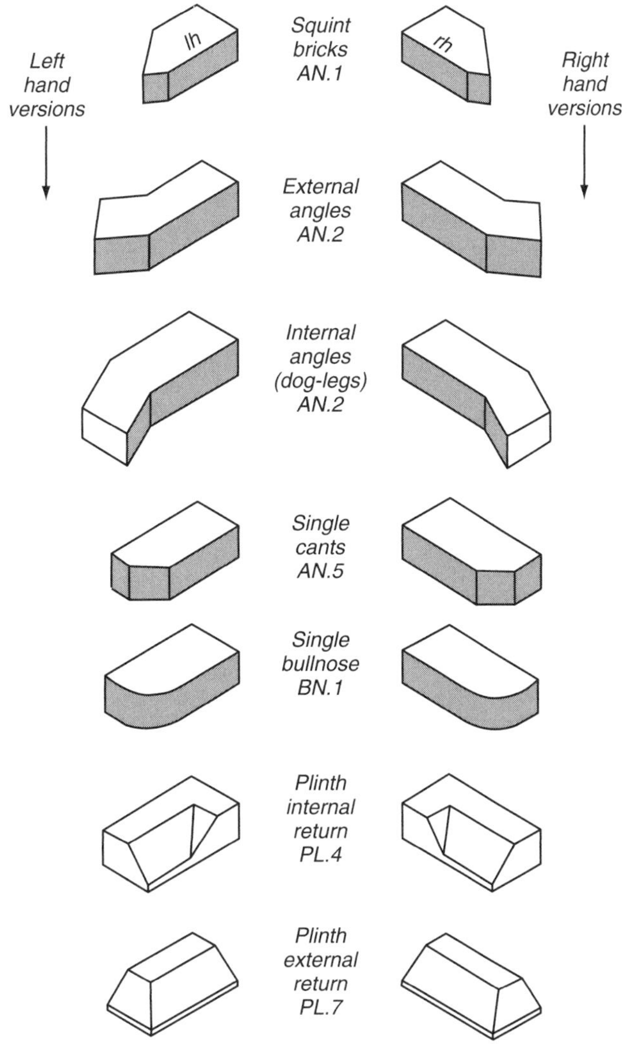

Figure 14.10 Handing of bricks of special shapes and sizes; BS 4729:1990 type numbers are included

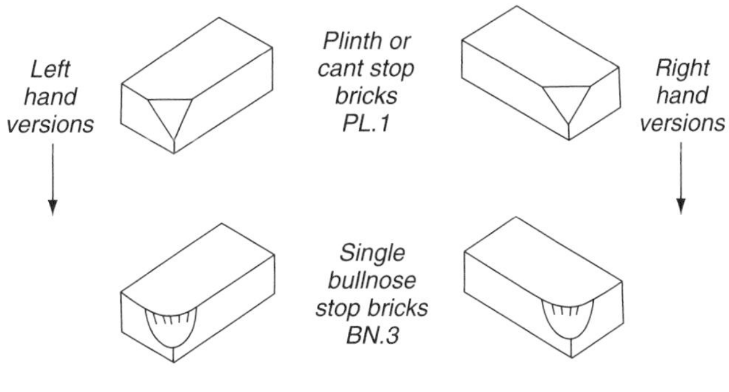

Left hand versions

Right hand versions

Squint bricks AN.1

External angles AN.2

Internal angles (dog-legs) AN.2

Single cants AN.5

Single bullnose BN.1

Plinth internal return PL.4

Plinth external return PL.7

Figure 14.11 Handing of special shape stop bricks

Left hand versions

Right hand versions

Plinth or cant stop bricks PL.1

Single bullnose stop bricks BN.3

Radial specials

In a similar way to arch bricks, the BS details some commonly used radial bricks with type numbers that can be quoted when ordering, see Fig. 14.12. If a different radius of curvature is required to construct curved brickwork, then detailed drawings must be sent to the brick manufacturer, from which special moulds can be made.

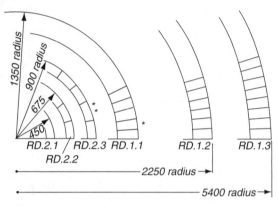

Figure 14.12 Radial bricks. Plan view showing six standard radius curves, all available in BS 4729 as either tapered headers or tapered stretchers

**Extrados dimension 226 mm
*Extrados dimension 108 mm

Plan views of walling

15   Joint finishing

The surface finishing treatment of new face work may be a jointing or pointing operation for bricklayers, and has an important effect upon the finished appearance of brickwork. When looking at stretcher bonded face work 18% of what you see is mortar colour.

Jointing is the craft term applied when joints are finished with the same mortar as is being used for the bricklaying, while the work proceeds.

Pointing is the term used to describe the surface finish applied to the cross joints and bed joints of a brick wall when raked out to a depth of approximately 12 mm, and filled with a mortar of different colour or texture.

Figures 15.1(a) and 15.1(b) show a number of different joint finish profiles.

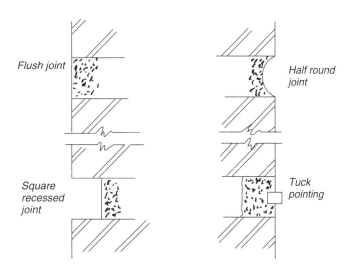

Figure 15.1(a) Joint types

Figure 15.1(b) Joint types

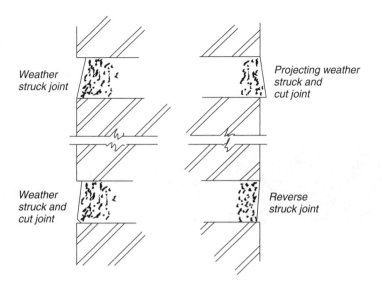

Weather struck joint

Projecting weather struck and cut joint

Weather struck and cut joint

Reverse struck joint

Reasons for joint finishing

The purpose of joint finishing is to compact the surface and press the mortar into contact with the arrises of each brick, so as to prevent rain penetration and possible frost damage.

Another reason for jointing or pointing is to give the exposed mortar surfaces an even appearance overall. The word 'weather' used in any description of jointing or pointing simply means 'sloped', so as to shed or tip rainwater off the surface of joints.

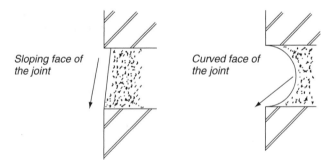

Sloping face of the joint

Curved face of the joint

Figure 15.2 Weathering of joints

Mortar mixes

Where face work is to be 'jointed as work proceeds', the bricklaying mortar of course provides the joint finishing colour. 'Ironing in' bricklaying mortar made from a fine grain building sand, for example, will leave a smoother surface than where coarser local sand is in use.

If fine grain building sand is used to produce pointing mortar, then a 'weather struck and cut' finish will polish up better and may be 'cut' or trimmed more cleanly with the frenchman, than when the sand is coarser.

Cement-rich or 'strong' pointing mortar, e.g. BS 3921 Group (i), should be reserved for very dense Class 'A' engineering bricks only. Group (ii) $1:\frac{1}{2}:4\frac{1}{2}$ is suited to Class 'B' bricks.

A slightly stronger mix than Group (iii) 1:1:6 is in order for the majority of bricks with compressive strengths between 20 and 40 N/mm^2, e.g. 1:1:5. The reduction in sand improves the 'fattiness' of the mortar, so that it sticks to the pointing trowel. If the fattiness of a pointing mortar needs further improvement, it is better to increase the proportion of lime rather than the cement.

Very careful and consistent batching is necessary, with strict control of mix proportions by volume using gauge boxes or buckets each time, if mortar is always to finish up the same colour and strength.

Sample panels

Before face work commences, small sample walls of approximately 1 m should be built and jointed or pointed, so that the architect may see the effect of the specified joint finish and its colour. More than one sample panel may be required and each should be allowed to dry out fully before final choices are made.

Tools

Those items in a bricklayer's tool kit used for joint finishing can be a mixture of purchased and home-made items as shown in Fig. 15.3.

Choice of joint finish

Flush joints

Flush jointing is usually specified when the architect does not wish to see any shadow lines cast by the joint finish. Another reason for its use is where a 'matt' joint surface is preferred, instead of one smoothed or struck by a steel trowel or metal jointer.

Although this is a very simple looking joint finish, it is not as easy as it looks to get the mortar surface truly flush – not slightly 'dished'. The hardwood block or PVC slip shown in Fig. 15.3 must be used to smooth and compact the surface mortar, but must not leave it pressed further back than the face of the bricks; see 'Technique – jointing' below.

Flush jointing with hand-made or stock bricks can make joints appear wider than they really are, due to the rounded arrises of these bricks.

Weather struck and cut pointing

This can make courses of irregular-shaped bricks appear straighter than they are; see 'Technique – pointing' below.

Possibly the most commonly specified joint finish where new brickwork has been raked out, and is to be pointed at a later date or when old brickwork needs re-pointing. This joint finish ensures that rain is tipped

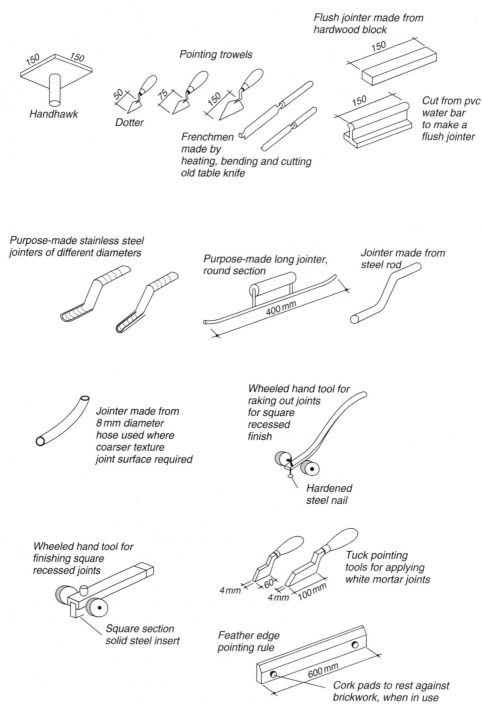

Figure 15.3 Tools for jointing and pointing

off the exposed bed joint surfaces and arris of each brick; but it is not practical to use as a 'jointing' operation as work proceeds. The additional mortar which must be applied to permit neat 'cutting' of cross and bed joints (see Figs 15.5 and 15.6) would delay economical bricklaying progress.

Weather struck joints

This 'weathered' joint finish can be carried out as a jointing operation while work proceeds, for no delaying 'cutting' is required. It includes the advantages of a sloping surface to the exposed bed joints, to improve weather resistance.

The depth of indenting the left-hand side of cross joints and the top of bed joints of this profile should not be deeper than the trowel blade thickness. Care must be taken to see that both left-handed and right-handed bricklayers indent cross joints on the LHS only.

Bricklayers develop an individual style when performing this jointing operation; some 'strike' the mortar joints as they lay each brick, others complete a whole course and then proceed to strike up the joints. In either case, it is advisable to use the brick trowel rather than a pointing trowel for struck jointing.

Reverse struck joints

This joint finish is used for internal wall surfaces of common brickwork that are to be left as a finish or coated with emulsion paint. Joints are struck the opposite way to weather jointing, and not trimmed or 'cut'. Walls have a flat even appearance when viewed from below, free of joint shadows. This trowelled finish performed with the bricklaying trowel rather than a pointing trowel, should be as flat as possible to avoid ledges which act as dust traps.

Ironed or tooled joints

This is the most common joint finish, allowing the bricklayer to disguise slight chips or imperfections on the arrises of bricks so as to avoid unnecessary waste.

Care should be exercised to see that all the bricklayers in a gang use the same diameter jointers. Smaller diameters give a deeper joint profile. Larger diameter jointers give a shallower profile.

Irregular depth ironing of joints will cast different shadow effects, and cause variations in the appearance of completed elevations of face work. (See 'Technique – jointing' below.)

Square-recessed jointing

This finish gives a strong shadow effect, but exposes every chip or imperfection on the arrises of bricks. As an external joint finish must only be used with frost resistant bricks, because bed joints are not weathered, rain can lodge on recessed arrises giving the risk of frost damage.

The depth gauge on each bricklayer's wheeled jointer must be set the same, e.g. 6 mm or 8 mm, when raking out mortar joints. All square-recessed joints must then be polished directly with the solid steel insert of the wheeled jointer to leave the recessed mortar surfaces smooth, square and clean; see Fig. 15.3.

Tuck pointing

A now practically obsolete and extremely time consuming system of pointing which applies false cross joints and false bed joints only 3 mm wide to the face of previously flush jointed brickwork. It developed in eighteenth century Britain as a next best thing, if you could not afford genuine gauged brickwork built with red rubbers and fine white lime putty joints. In modern practice it is generally only used where there is a requirement to match existing work during refurbishment of period buildings.

The false white mortar joints, referred to as 'the strip' by the few modern tuck pointers, is superimposed on a flush joint, called 'the stopping'.

Stopping mortar for the flush joint is either coloured to suit the bricks in the wall, or in the case of old buildings, ordinary mortar is used, and the whole wall face is colour washed before 'the strip' is applied.

Tuck pointing mortar consists of a mix of lime putty and silver sand 1:3, although modern tuck pointers frequently include $1/2$ part of white cement to make the strip more durable. The flush pointed stopping mortar should be keyed or grooved on the previous day so that the white tuck pointing is firmly attached (see Fig. 15.1(a)).

Gauge rods are necessary to ensure exact spacing of these joint lines marked in the flush stopping mortar, and the bond pattern accentuated with perfectly plumb perpends.

The white mortar is loaded carefully along one edge of the pointing rule, from which it is pressed on to the face of the flush joint stopping, using purpose-made tuck pointing tools (see Fig. 15.3).

Bed joints must be applied first (the opposite way round compared to weather struck and cut pointing), and trimmed top and bottom using a feather edge pointing rule and frenchman. Cross joints are applied last and similarly trimmed with pointing rule and frenchman.

Technique – jointing

Whichever finish is to be used, timing is very important when jointing-up as the bricks are laid throughout the working day. The mortar between bricks should be allowed to stiffen up just enough due to brick suction, so that the jointing tool can pass smoothly and cleanly. Too soon and the mortar smears and does not leave a smooth profile. Left too long before jointing and heavy pressure on the jointer leaves black metal marks on the dried mortar face.

With the single exception of tuck pointing, whether jointing or pointing, always do the cross joints first, followed by bed joints, each time you stop bricklaying to joint up.

Brushing off with a soft bristle hand brush, to remove any loose crumbs of mortar, should be left until the end of the day. Brush lightly and on no account leave bristle marks in the mortar face. Better to leave brushing until the following morning than to risk marking the joints. Take particular care when jointing face brickwork at those points shown in Fig. 15.4.

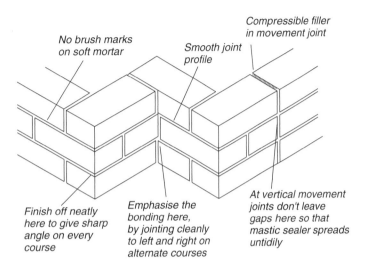

No brush marks on soft mortar

Smooth joint profile

Compressible filler in movement joint

Finish off neatly here to give sharp angle on every course

Emphasise the bonding here, by jointing cleanly to left and right on alternate courses

At vertical movement joints don't leave gaps here so that mastic sealer spreads untidily

Figure 15.4 Fine points of jointing and pointing

Technique – pointing

This craft operation, carried out some weeks or months after the wall has been built, requires patience and is a skill which takes time to develop. The joint finish commonly specified for pointing brickwork is weather struck and cut (see Fig. 15.1b).

1. Always start at the very top of the walling to be pointed.
2. Remove any obvious hardened crumbs of mortar clinging to the wall face when the joints were raked out some weeks or months ago.
3. Brush the whole lift of brickwork using a stiff bristle hand brush.
4. Wet the wall face generously if the bricks are very absorbent, less generously if the bricks have a lower suction rate. (See item 14 below for walls with alternate bands of Class 'A' and absorbent facings.)
5. Load the hand hawk with mortar flattened out to approximately 10 mm thickness.
6. Using the small pointing trowel or 'dotter', pick up joint sized pieces of mortar from the hawk and press carefully and firmly into cross joints, but see also item 14 below.
7. Completely fill each cross joint with a second application if necessary, polish the mortar surface and indent on the LHS.
8. After completing approximately 1/2 m² of cross joints, cut or trim the RHS of all these joints in the manner shown in Fig. 15.5, so that all look the same width on face.

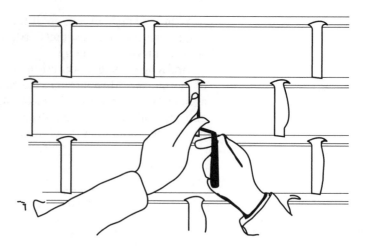

Figure 15.5 Weather struck and cut pointing: cutting the right-hand side of cross joints

9. Using a longer pointing trowel, pick up joint sized pieces of mortar from the hawk and commence pointing one bed joint, pulling the loaded trowel up to the last piece of mortar applied each time.
10. After filling a 500 mm length of bed joint, polish it with the pointing trowel and indent the top.
11. When half a dozen bed joints have been pointed in this way, cut or trim the bottom edge of each one as shown in Fig. 15.6 using frenchman or tip of pointing trowel together with feather edge pointing rule.

   The amount of mortar to be left projecting from the wall face after trimming or cutting joints should not exceed thickness of the trowel blade.

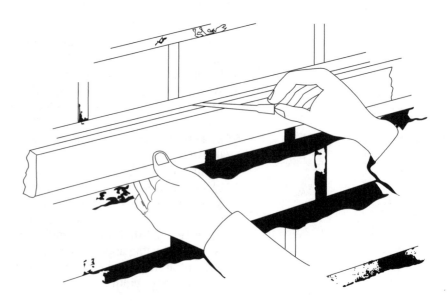

Figure 15.6 Weather struck and cut pointing: trimming the bottom of bed joints with straight edge and frenchman

12. Sensibly adjust areas of pointing to suit drying conditions of bricks and weather, so that joints cut cleanly. Too early, and the mortar will not fall away cleanly when trimmed. Too late, and joint edges will crumble and not leave a clean straight line when cut.
13. Brush very lightly at the end of the day or preferably on the following day if there is the slightest risk of marking the sharp cut edges of the pointing.
14. If a wall has alternate bands of Class 'A' and absorbent facing bricks, after wetting the whole wall, point absorbent facings first. When the wall has dried off, return and point the Class 'A' bands.

Re-pointing old brickwork

Chimney stacks and parapet walls are usually the first parts of a brick building which will need re-pointing after twenty or thirty years exposed to wind, rain and frost. Raking out and re-pointing can give such brickwork a new lease of life, providing that it is otherwise structurally sound.

1. Rake out all old mortar, using bolster and comb chisel, to a depth of 15–18 mm, taking care not to damage brick arrises unnecessarily.
2. Remove all old mortar and dust by vigorously brushing brickwork with a stiff bristle brush.
3. Thoroughly wet the brickwork, and when surface is dry, commence re-pointing from the top down.
4. Continue as outlined in this chapter, sub-heading 'Technique – pointing'.

Summary

The great merit of surface finishing mortar as a 'jointing' process is that the joint profile is an integral part of the mortar bed and there is no possibility of failure through insufficient adhesion between the main mortar bed and the surface finish.

Failure of pointing, i.e. its separation from the main mortar bed and consequent falling away, is caused by careless raking out and lack of suitable preparation. To overcome possible failure, joints in new brickwork should be raked out to a depth of at least 12 mm as shown in Fig. 15.7 and not as shown in Fig. 15.8.

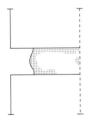

Figure 15.7 Correct way of raking out joints in preparation for re-pointing

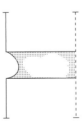

Figure 15.8 Incorrect way of raking out, which will lead to pointing failure due to poor adhesion

16 Calculations

A bricklayer should know how to work out the number of bricks needed to build a wall. This is a very simple matter using a pocket calculator. Do not panic at the very mention of arithmetic. Let the calculator take the strain and deal with the decimal point!

Method A

There is nothing wrong with counting up the number of bricks needed for one course – then multiplying by the total number of courses in the wall to get the answer.

This method works well for stretcher bond walling, but becomes difficult when applied to more complicated shapes that have doors and windows and also with cavity walling.

Method B

This method is based on finding the surface area of the wall. It makes deductions for doors and windows easy. Use the calculator to add, subtract and multiply. Do not strain the brain!

Types of measurement

Below are the basic types of measurement which the bricklayer will have to understand.

Number

This is used for counting how many of a particular item are required. For example, how many windows, doors or air bricks are required.

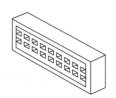

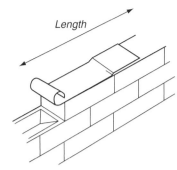

Linear measurement

This is used to measure the length of a particular item such as the length of copings required for a wall.

Area

This is a superficial measurement used for calculating areas. Two measurements are required to be multiplied together such as the length and width of a piece of ground.

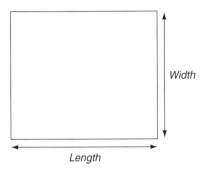

Volume

This calculation requires three figures to calculate the volume and is also known as a cubic measurement.

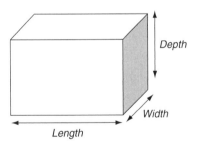

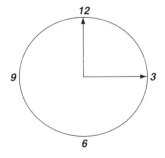

Time

Time is used by all firms when calculating bills using information from time sheets.

Cost

Estimating the cost of a job is impossible without knowledge of money. All employees in the construction industry are employed on a certain

rate for the job. For example, the current rate for a bricklayer may be £5.80 per hour. This involves a calculation involving both time and money.

Remember

All calculations should be shown and not done on scrap bits of paper or in your head. This will enable any mistakes to be found. When end tests are being carried out marks are given for the correct method of setting out and not just for the answer.

Calculators

These are a very useful tool. BUT do not rely on the calculator at first. A basic calculator with $+$, $-$, $\times$, $\div$, is all that is needed. Memory in a calculator can also be useful, but all steps need to be shown on paper. It is particularly important to check answers when using calculators because it is too easy to press the wrong button on small calculators.

Decimals

It is important that the decimal points are correctly placed. Misplacing a decimal point can make a result ten times too large or small.

The metric system

The metric system has been devised so that standard units are used throughout the world. The metric system is much easier to use in calculations and measurements.

Unit and length	Symbol
Metre	m
Millimetre	mm
Kilometre	km

1 km = 1000 m, 1 m = 1000 mm

Therefore 1 km = 1 000 000 mm

Unit of mass	Symbol
Kilogram	kg
Gram	g
Tonne	t

1 Tonne = 1000 kg

1 kg = 1000 g

Unit of force	Symbol
Newton	N
Kilonewton	kN
Meganewton	MN

1 MN = 1000 kN

1 kN = 1000 N

To convert mass (weight) to force = multiply by 9.81 m/s.

Brick calculations

The size of a brick including the joints = 225 × 112.5 × 75 mm.

If we find the face area of one brick and divide that into 1 square metre it will result in the brick requirement.

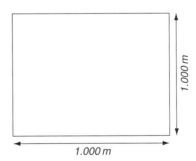

Area of the stretcher face of a brick $0.225 \times 0.075 = 0.01687$ m^2
Area of a half brick wall = $1 \times 1 = 1$ m^2.

Therefore

$$\frac{1}{0.01687} = 59.27 \text{ bricks}$$

Allowing for waste = 60 bricks per metre square of wall half brick thick. Other brick requirements are:

1 brick wall	120 bricks
$1^1/_2$ brick wall	180 bricks

Mortar calculations

Nominal size for a brick is $215 \times 102.5 \times 65$ mm.
Nominal size for a joint is 10 mm.

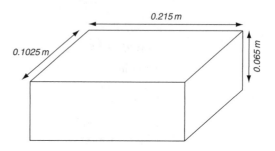

If the brick size = $215 \times 102.5 \times 65$ mm then the bed joint
$= $ length $\times$ width $\times$ thickness of bed joint
$= 0.125 \times 0.1025 \times 0.01$
$= 0.00013$ m^3

the cross joint $= $ width $\times$ depth $\times$ thickness of cross joint
$= 0.1025 \times 0.075 \times 0.01$
$= 0.00008$ m^3.

Therefore the total for one bed joint and one cross joint
$= 0.00013 + 0.00008$
$= 0.0002$ m^3.

If the calculation was for 1000 bricks the mortar required
$= 0.0002 \times 1000$
$= 0.2$ m^3.

Note: Normal allowance is around 0.6 m^3 per 1000 bricks which allows for the frogs and waste. Another method to simplify the calculation is to allow 1 kg of mortar for each brick.

It follows that 1000 bricks would require 1000 kg of mortar which is 1 tonne.

Therefore $1/_2$ brick wall = 60 bricks + 60 kg mortar
1   brick wall = 120 bricks + 120 kg mortar
$1^1/_2$ brick wall = 180 bricks + 180 kg mortar

Example 1

Calculate the number of bricks required for the wall area.

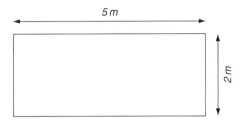

Answer 1

Surface area = 5 × 2 = 10 m² (square metres)

You require 60 bricks (from previous calculations) to build one m² of $^1/_2$ brick thick (102 mm) stretcher bond walling.

The surface area of the wall is 10 m², then you need 10 × 60 = 600 bricks.

Example 2

Calculate the number of bricks required for the wall area.

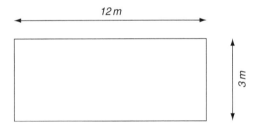

Answer 2

Surface area = 12 × 3 = 36 m² (square metres)

You require 60 bricks to build one m² of $^1/_2$ brick thick (102 mm) stretcher bond walling.

The surface area of the wall is 36 m², then you need 36 × 60 = 2160 bricks.

STOP

When doing any calculation always write down your working method as shown by the lines in Example 2. This is worth it, because you can then check what you did. It will also ensure that you do not get in a muddle with the decimal point.

Walls 1 B thick (215 mm) need 120 bricks to build one m² (1 m²), and the same working method can be used as shown in Example 2.

Example 3

How many bricks are required to build this wall?

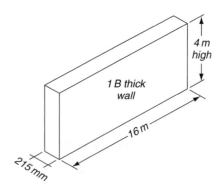

Answer 3

120 bricks per m²
Two lines of working method:

Surface area $= 16 \times 4 = 64$ m²

Number of bricks needed $= 64 \times 120 = \underline{7680 \text{ bricks}}$.

Walls are not always straight elevations, and buildings have corners, but the same method of calculation can be used.

Example 4

How many bricks are required to build the wall shown?

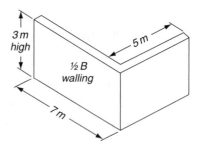

Answer 4

60 bricks per m²
Three lines of working method here:

Total length of wall $= 7 + 5 = 12$ m
Surface area $= 12 \times 3 = 36$ m²

Number of bricks required $= 36 \times 60 = \underline{2160 \text{ bricks}}$.

STOP

If calculations are not a happy thought for you, it is always a good idea to retrace your steps. Go back to Example 1. Just change the dimensions of that wall, invent your own figures, and then work out how many bricks will be needed this time.

Repeat what you have just done, but with Examples 2, 3 and 4, to calculate the number of bricks required for each example, but with different wall dimensions.

Example 5

How many bricks are required to build this wall?

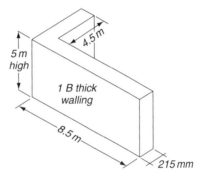

Answer 5

120 bricks per m²
Three lines of working method:

Total length = 8.5 + 4.5 = 13 m
Surface area = 13 × 5 = 65 m²

Number of bricks required = 65 × 120 = <u>7800 bricks</u>.

Example 6

How many bricks are required to build this wall?

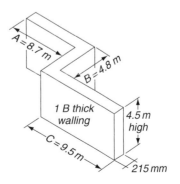

Answer 6

120 bricks per m^2
Three lines of working method:

Total length = A + B + C = 8.7 + 4.8 + 9.5 = 23 m
Surface area = 23 × 4.5 = 103.5 m^2

Number of bricks required = 103.5 × 120 = 12 420 bricks.

STOP

For additional practice, repeat Examples 5 and 6 after changing the dimensions, invent your own, and assume that the walls are only 102 mm thick.

Blockwork

A similar method can be used to find out how many blocks are needed to build a wall. The size of a block including the joints = 450 × 112.5 × 225 mm.

If we find the face area of one brick and divide that into 1 square metre it will result in the brick requirement.

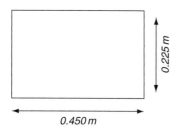

0.225 m

0.450 m

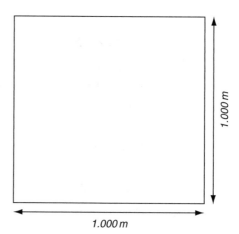

1.000 m

1.000 m

Area of the face of a block 0.450 × 0.225 = 0.10125 m^2
Area of a half brick wall = 1 × 1 = 1 m^2.

Therefore

$$\frac{1}{0.10125} = 9.88 \text{ blocks}$$

Allowing for waste = 10 blocks per metre square of wall.

Example 7

Calculate how many standard size blocks are required to build this wall.

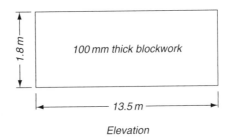

Elevation

Answer 7

10 blocks per m² (from previous calculations)
Two lines of working method:

Surface area = 13.5 × 1.8 = 24.3 m²
Number of blocks = 24.3 × 10 = <u>243 blocks</u>.

Example 8

How many standard size blocks are needed to build this wall?

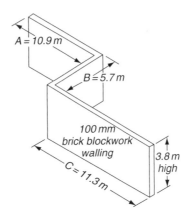

Answer 8

10 blocks per m²
Three lines of working method:

Total length of wall = A + B + C = 10.9 + 5.7 + 11.3 = 27.9 m
Surface area = 27.9 × 3.8 = 106.02 m²

Number of blocks = 106.02 × 10 = 1060.2
Rounded off to <u>1061 blocks</u>.

STOP

When ordering blocks for the inner leaf of cavity walls or for partition walls, take care to make clear the thickness you want, 100, 140, 150, 190 or 215 mm, as well as type, compressive strength, density and surface finish.

Mortar

Allow 0.02 m³ (cubic metre) (from previous calculations) of ready-mixed mortar to build one m² (1 m²) of stretcher bonded brickwork. This is an average figure, bearing in mind that some bricks have frogs and others are perforated.

Table 16.1 shows approximately how much ready-mixed mortar is required for different wall thicknesses in brick and block.

Example 9

Calculate how much mortar will be required to build the brick wall.

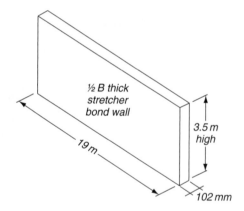

½ B thick
stretcher
bond wall

3.5 m
high

19 m

102 mm

Answer 9

0.02 m³ of mortar per m² of walling (from previous calculations)
Two lines of working method:

Surface area of wall = 19 × 3.5 = 66.5 m²

Volume of mortar required = 66.5 × 0.02 = <u>1.33 m³</u>.

STOP

For additional practice, go back to calculate how much mortar would be required to build the wall in Example 4.

Table 16.1 Materials per square metre of walling

	Stretcher bond			English bond			Flemish bond			English garden wall bond			Flemish garden wall bond			Header bond		
	F	C	M(m³)	F	C	M(m³)	F	C	M(m³)	F	C	M(m³)	F	C	M(m³)	F	C	M(m³)
½ B 102 mm thick	60		.02													120		.047
1 B 215 mm thick	60*	60*	.05	90	30	.05	80	40	.05	73†	47	.05	67†	53	.05			
1½ B 327 mm thick				90	90	.08	80	100	.08	73	107	.08	67	113	.08			
2 B 440 mm thick				90	150	.11	80	160	.11	73	167	.11	67	173	.11			
100 mm standard size blocks	10		.01															
140 mm standard size blocks	10		.014															
150 mm standard size blocks	10		.015															
190 mm standard size blocks	10		.019															
215 mm standard size blocks	10		.022															

*215 mm thick walls can be built to show stretcher bond and give a face finish if required both sides if steel mesh bed joint reinforcement or butterfly wire ties are incorporated every 4th course.

†Garden wall bonds are intended for 215 mm thick free standing boundary walls using 100% facing bricks.

Note: F = facings; C = commons; M = mortar.

Example 10

Calculate how many bricks and how much mortar will be required to build the following wall.

120 bricks and 0.05 m³ of mortar per m²

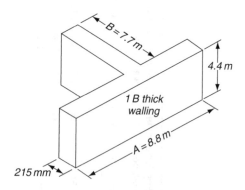

Answer 10

Four lines of working method:

Total length = A + B = 8.8 + 7.7 = 16.5 m
Surface area = 16.5 × 4.4 = 72.6 m²
Number of bricks = 72.6 × 120 = 8712 bricks

Volume of mortar = 72.6 × 0.05 = <u>3.63 m³</u>.

Example 11

Calculate how many standard sized blocks and how much mortar will be required to build the following block wall.

10 blocks and 0.01 m³ of mortar per m²

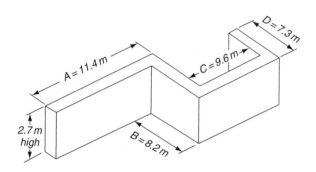

Answer 11

Four lines of working method:

Total length = A + B + C + D = 11.4 + 8.2 + 9.6 + 7.3 = 36.5 m
Surface area = 36.5 × 2.7 = 98.55 m²
Number of blocks = 98.55 × 10 = 985.5 blocks: <u>Rounded off to 986 blocks</u>

Volume of mortar = 98.55 × 0.01 = 0.9855 m³: <u>Rounded off to 1 m³.</u>

Rounding off

With building calculations, it is customary to 'round off' decimal parts, so as to tidy up quantities of materials into whole numbers where possible. For example, it is not practical to order half a bag of cement, nor worry about 0.2 of a cubic metre of a bulk item like trench excavation.

It is sensible to 'round up' with building materials, as site handling can result in damage or loss even in excess of the percentage allowances for cutting and waste. Taking Example 11, 1 m³ (1 cubic metre) is a less cumbersome figure than 0.9855 m³ and involves fewer numbers with which to make mistakes.

Openings

Most walls and buildings have door and window openings. This means that you must reduce your order for bricks and blocks, depending upon the size and the number of openings.

The best way is to deduct the surface area of an opening from the overall wall area, before the final step of calculating the number of bricks or blocks needed (see Example 12).

Example 12

Calculate how many bricks will be required to build this wall, making the necessary deduction for the opening.

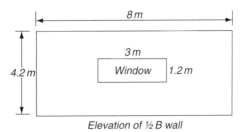

Elevation of ½ B wall

Answer 12

Using 60 bricks per m^2
Four lines of working method:

Overall surface area of the wall $= 8 \times 4.2 = 33.6$ m^2
Area of window opening $= 3 \times 1.2 = 3.6$ m^2

Nett area of walling $= 33.6 - 3.6 = 30$ m^2

Number of bricks required $= 30 \times 60 = \underline{1800 \text{ bricks}}$.

Example 13

How many bricks are required to build the wall shown, making allowances for openings?

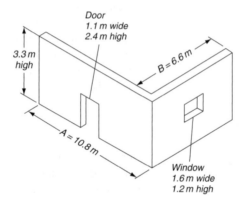

Door
1.1 m wide
2.4 m high

3.3 m
high

B = 6.6 m

A = 10.8 m

Window
1.6 m wide
1.2 m high

Answer 13

Using 60 bricks per m^2
Five lines of working method:

Total length of wall $= A + B = 10.8 + 6.6 = 17.4$ m

Surface area $= 17.4 \times 3.3 = 57.42$ m^2
Area of door opening $= 1.1 \times 2.4 = 2.64$ m^2
Area of window opening $= 1.6 \times 1.2 = 1.92$ m^2

Nett area of walling $= 57.42 - 4.56 = 52.86$ m^2

Number of bricks required $= 52.86 \times 60 = 3171.6$ bricks

Rounded off to $\underline{3172 \text{ bricks}}$.

STOP

For additional practice, calculate how much ready-mixed mortar will be required to build each wall shown in Examples 12 and 13. (Follow lines of working method given in Example 9.)

Cavity walling

When working out quantities of materials for cavity work, each leaf of the walling must be dealt with on its own, because you need separate totals for ordering bricks and blocks (see Example 14).

Example 14

Calculate how many facing bricks and blocks will be required to build the cavity shown, making due allowances for the opening.

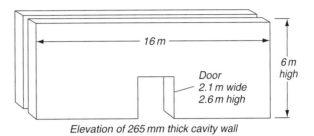

Door
2.1 m wide
2.6 m high

16 m

6 m
high

Elevation of 265 mm thick cavity wall

Answer 14

Using 60 bricks and 10 blocks per m^2
Five lines of working method:

Overall surface area $= 16 \times 6 = 96 \ m^2$
Area of opening $= 2.1 \times 2.6 = 5.46 \ m^2$
Nett area of walling $= 96 - 5.46 = 90.54 \ m^2$

Number of facing bricks $= 90.54 \times 60 = 5432.4$
Rounded off to 5433 bricks

Number of blocks $= 90.54 \times 10 = 905.4$
Rounded off to 906 blocks.

Perimeters

Perimeter of a rectangle $= 2 \times$ length $+ 2 \times$ width.

Example 15

Find the perimeter of the area shown.

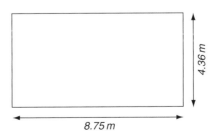

4.36 m

8.75 m

Answer 15

Perimeter = 2 × L + 2 × W
          = (2 × 8.75) + (2 × 4.36)
          = 17.50 + 8.72
          = <u>26.22 m</u>.

Example 16

Find the perimeter of the area shown.

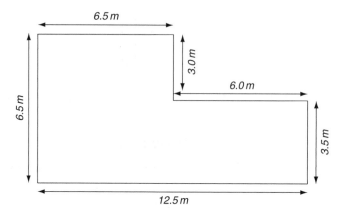

Answer 16

Perimeter = 2 L × 2 W
          = (2 × 12.5) + (2 × 6.5)
          = 25 + 13
          = <u>38 m</u>.

There is no need to add all the individual sides when the overall length and width is known. If the dimension was missing from the long side then the two smaller sides would have to be added together to give the overall length = 6.5 + 6 = 12.5 m.

STOP

The reason for deducting or adding 4 × the width of the walls when calculating the centre line is as follows. Irrespective of the design of a building there will always be 4 external right angles. Check out the following designs.

Example 17

The first design shows 5 external corners and 1 internal. If you deduct the internal from external corners you are left with 4 external corners. This rule applies to all rectangular designs.

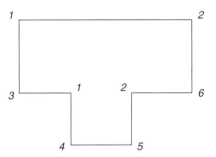

Example 18

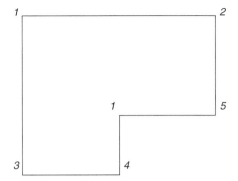

The second design shows 6 external corners and 2 internal. If you deduct the internal from external corners you are still left with 4 external corners. This rule applies to all rectangular designs with corner insets.

Remember also that the perimeters of designs with the same overall dimensions are the same irrespective of the number of corner insets.

Example 19

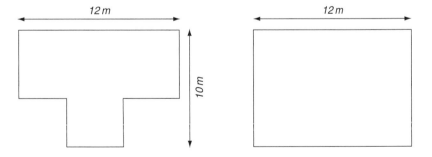

The two designs have the same perimeters even if the shapes are different.

The only time the perimeters are different and the number of external angles are more than 4 is when the design incorporates insets.

Example 20

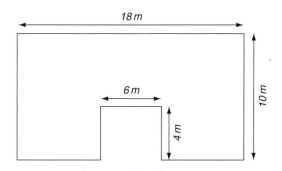

Answer 20

The perimeter is equal to that of a rectangle with sides of 18 and 10 plus the depth of each insert line.

Therefore: perimeter $= (2 \times 18) + (2 \times 10) + (2 \times 4) = 64$ m.

Method C.
The centre line method

When walls have many more returns than Example 11, it can be difficult to ensure that corners are not measured twice over. Professional quantity surveyors make use of the 'Centre Line Method' to get over this problem when measuring up walls.

The previous two examples explained how the external perimeter of a building is calculated, but it is necessary to move this line into the centre of each wall to ensure the calculation is correct. This is done by making a deduction for each external angle equivalent to the thickness of each wall. If the dimensions were internal then an addition could be made for each internal angle equal to the thickness of each wall.

Example 21

Calculate the centre line for the building with a solid brick wall.

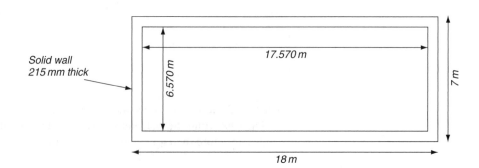

Answer 21

Taking external dimensions

Total external dimensions $= (18 + 7) \times 2 = 50$ m
Deduct corners $= 4 \times 215 = 0.860$ m

Centre line of solid brick wall $= 50 - 0.860 = \underline{49.140 \text{ m}}$

Taking internal dimensions

Total internal dimensions $= (17.570 + 6.570) \times 2 = 48.280$ m
Add corners $= 4 \times 215 = 0.860$ m

Centre line of solid brick wall $= 48.280 + 0.860 = \underline{49.140}$ m.

The following enlarged drawing will explain in more detail why it is necessary to either add or deduct the full thickness when adjusting for corners.

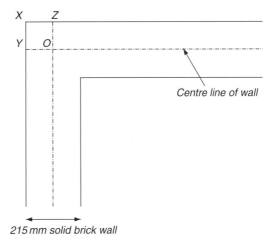

215 mm solid brick wall

The drawing shows one corner of a 215 mm solid brick wall. The centre line passes through the intersection point at 'O'.

The centre lines both extend to meet the outside face of the wall at 'Y' and 'Z'. This shows quite clearly that the lines 'XY' and 'XZ' need to be deducted to produce the correct centre line.

The two dimensions are equivalent to two half walls which together add up to a full wall thickness. There are always four corners which need either deducting when taking external dimensions or adding when taking internal dimensions.

When using the centre line method on solid walls it gives the centre line for the excavation, the foundation concrete, the damp proof course and the brickwork. So you can see it is a very important calculation.

When cavity wall construction is being measured the calculation is a little more complicated but the same principle applies.

The centre line of the wall will be required for the excavation, foundation concrete and the formation of the cavity and possible cavity insulation, but a deduction or addition will need to be made to calculate the centre line of the external facings and the inner blockwork.

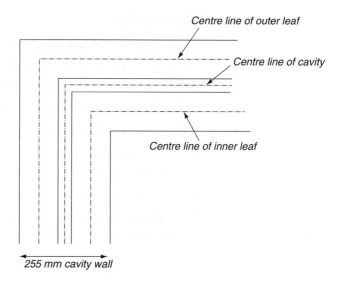

Centre line of outer leaf

Centre line of cavity

Centre line of inner leaf

255 mm cavity wall

Assume the external wall to be 102.5 mm wide, the cavity 50 mm and the inner blockwork 102.5 mm.

To arrive at the centre line of the face brickwork, taking external dimensions, you would need to deduct 4 × 102.5 from the overall perimeter dimensions. Then to calculate the centre line of the cavity you first need to deduct the same figure (4 × 102.5) to arrive at the inside face of the outer leaf of face brickwork. Then deduct 4 × 50 to find the centre line of the cavity.

To find the centre line of the blockwork you first need to move the centre line to the inner face of the blockwork by deducting 4 × 50 mm cavity. Deduct 4 × 102.5 to find the centre line of the blockwork.

Example 22

Calculate the centre line for both the outer leaf of facing bricks and the inner leaf of blocks.

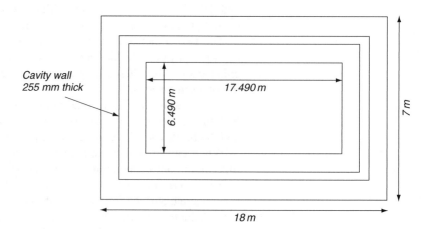

Cavity wall 255 mm thick

6.490 m

17.490 m

7 m

18 m

Answer 22a: External dimensions

Centre line of outer leaf of facing bricks:

Total external dimensions = $(18 + 7) \times 2 = 50$ m
Deduct corners = $4 \times 102.5 = 0.410$ m

Centre line of facing brick outer leaf = $50 - 0.410 = \underline{49.59 \text{ m}}$

Centre line of inner face of facing bricks:

Centre line of outer leaf = 49.59 m
Deduct $4 \times 102.5 = 0.410$ m
Inner face of brick outer leaf = $49.59 - 0.410 = \underline{49.18 \text{ m}}$

Centre of cavity:

Centre line of inner face of facing bricks = 49.18 m
Deduct $4 \times 50 = 0.200$ m
Centre line of cavity = $49.18 - 0.200 = \underline{48.98 \text{ m}}$

Inner face of blockwork:

Centre line of cavity = 48.98 m
Deduct $4 \times 50 = 0.200$ m

Centre line of inner face of blockwork = $48.89 - 0.200 = \underline{48.78 \text{ m}}$

Centre line of blockwork:

Centre line of inner face of blockwork = 48.78 m
Deduct $4 \times 102.5 = 0.410$ m
Centre line of inner leaf of blockwork = $48.78 - 0.410 = \underline{48.37 \text{ m}}$

Answer 22b: Internal dimensions

Centre line of inner leaf of blockwork:

Total internal dimensions = $(17.49 + 6.49) \times 2 = 47.96$ m
Add corners = $4 \times 102.5 = 0.410$ m

Centre line of blockwork inner leaf = $47.96 + 0.410 = \underline{48.37 \text{ m}}$

Centre line of inner face of blockwork:

Centre line of inner leaf = 48.37 m
Add $4 \times 102.5 = 0.410$ m
Inner face of blockwork inner leaf = $48.37 + 0.410 = \underline{48.78 \text{ m}}$

Centre of cavity:

Centre line of inner face of blockwork = 48.78 m
Add $4 \times 50 = 0.200$ m
Centre line of cavity = $48.78 - 0.200 = \underline{48.58 \text{ m}}$

Inner face of brickwork:

Centre line of cavity = 48.58 m

Add $4 \times 50 = 0.200$ m

Centre line of inner face of brickwork $= 48.59 + 0.200 = \underline{48.79 \text{ m}}$

Centre line of brickwork:

Centre line of inner face of brickwork $= 48.79$ m

Add $4 \times 102.5 = 0.410$ m

Centre line of outer leaf of brickwork $= 48.79 + 0.410 = \underline{49.20 \text{ m}}$

If you check the above answers they will be the same irrespective of which method is used.

If we combine the centre line method to the material calculation we should arrive at the total materials required for any building.

Example 23

Using the centre lines calculated in Example 22, calculate the facing bricks and blocks required for the building assuming the walls are 2.2 m high.

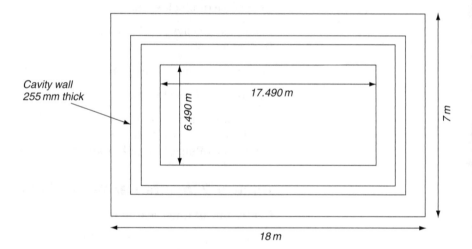

Answer 23

Using 60 bricks and 10 blocks per m²

Number of facing bricks:

Centre line of facing bricks $\times 2.2 = 49.20 \times 2.2 = 108.24$ m²

Number of facings $= 60 \times 108.24 = 6494.4$ bricks

Rounded off to <u>6494 bricks</u>

Number of blocks:

Centre line of blockwork $\times 2.2 = 48.37 \times 2.2 = 106.414$ m²

Number of blocks $= 10 \times 106.414 = 1064.14$ blocks

Rounded off to <u>1065 blocks</u>

Example 24

Calculate the total area for re-pointing the brick wall, back, front and ends.

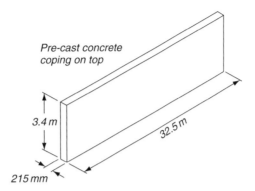

Answer 24

Perimeter = (32.5) × 2 + (0.215 × 2)
       = 65 + 0.43
       = 65.43 m

Total area of re-pointing = 65.43 × 3.4 = 222.462 m²

Rounded off to 233 m².

Example 25

Calculate the total area of raking out joints and re-pointing external walls of the building shown. (Ignore door and window openings for this example.)

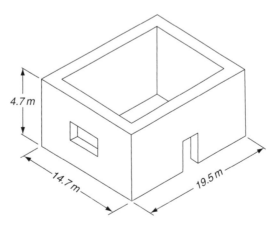

Answer 25

Perimeter = (19.5) × 2 + (14.7 × 2)
       = 39 + 29.4
       = 68.4 m

Total area of re-pointing = External perimeter × 4.7
                                       = 68.4 × 4.7
                                       = 321.48 m²

Rounded off to <u>322 m²</u>.

STOP

For extra practice repeat calculations in Examples 24 and 25, after changing the dimensions; invent your own.

Division of facings and commons

The outer, 102 mm thick leaf of cavity work and other ¹/₂ brick walling is usually built in stretcher bond, so that all the bricks required will be facings.

With 215 mm and thicker solid brickwork faced on one side only, some bricks will need to be less expensive commons to back up the facing bricks.

With 215 mm thick Flemish bond walling, ²/₃ will be facings and ¹/₃ commons. With 215 mm English bond walling, ³/₄ will be facings and ¹/₄ will be commons.

If 215 mm thick walls are built to header bond, then of course all bricks ordered should be facings. This information is summarised in Table 16.1.

Example 26

Assume that the wall shown is to be 215 mm thick Flemish bonded brickwork. Calculate the number of facings and commons that will be required, in separate totals.

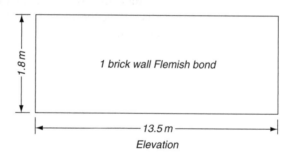

1 brick wall Flemish bond

1.8 m

13.5 m

Elevation

Answer 26

Taking information from Table 16.1

80 facings per m²
40 commons per m²
Three lines of working method:

Surface area = 13.5 × 1.8 = 24.3 m²

Number of facings = 24.3 × 80 = 1944 facing bricks

Number of commons = 24.3 × 40 = 972 common bricks.

Percentage for cutting and waste

All of the foregoing examples of calculating requirements for bricks, blocks and mortar have resulted in exact or nett quantities. Due to breakages and waste when cutting and handling bricks and blocks and when using mortar, it is necessary to increase nett orders by 5%.

Therefore as an additional stage at the end of each calculation, 5% extra should be added on, using the % button on your calculator, for the purposes of ordering.

Example 27

Calculate the amount of facing bricks and common bricks for the wall shown if English bond was used. Allow 5% for wastage.

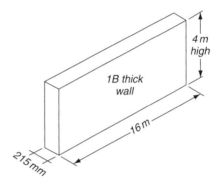

Answer 27

Taking information from Table 16.1

90 bricks per m²
30 commons per m²
Two lines of working method:

Surface area = 16 × 4 = 64 m²
Number of facings = 64 × 90 = 5760 facing bricks

Add 5% for wastage $= \dfrac{5760}{100} \times 5 = 288$

Therefore total facing bricks required = 5760 + 288 = <u>6048 facing bricks</u>

Number of commons = 64 × 30 = 1920 commons

Add 5% for wastage $= \dfrac{1920}{100} \times 5 = 96$

Therefore total commons required = 1920 + 96 = <u>2016 common bricks</u>.

STOP

Go back to Example 23 and calculate the number of bricks and blocks that should be ordered after allowing 5% for cutting and waste.

Example 28

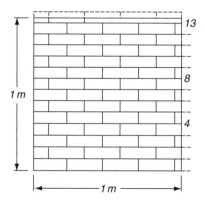

To calculate the area of mortar joints (say to assess the effect of mortar colour on finished walling) showing on the surface of one square metre of stretcher bonded brickwork.

For the purposes of this calculation, assume that the elevation above is 'stack bonded' then:

Answer 28

Bed joint mortar showing $= 1 \times 0.01 \times 13 = 0.13 \text{ m}^2$

Cross joint mortar showing $= 1 \times 0.01 \times 5 = 0.05 \text{ m}^2$

$$\text{Total} = 0.18 \text{ m}^2$$

$$\text{Expressed as a } \% \text{ of } 1 \text{ m}^2 = \frac{0.18}{1 \text{ m}^2} \times 100$$

$$= 18\% \text{ of surface viewed is mortar colour.}$$
The remaining 82% only is brick colour.

Additional applications of the centre line method

In addition to bricks, blocks and mortar, a bricklayer may be asked to work out quantities of cavity insulation, dpc, foundation concrete or even excavation to trenches.

Each such calculation is very simple, if you find out the total length of the centre line of the walling first.

Example 29

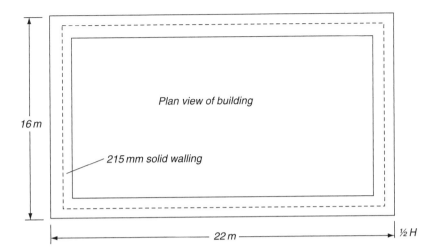

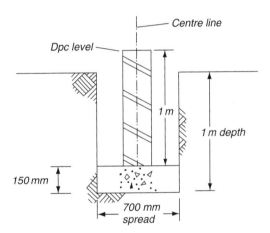

Answer 29

Centre line

Total external dimensions = $(22 + 16) \times 2 = 76$ m
Deduct corners = $4 \times 215 = 0.860$ m

Centre line of solid brick wall = $76 - 0.860 = \underline{75.14 \text{ m}}$.

The dotted centre line perimeter shown on the drawing is exactly 75.14 m long. This dotted line is the exact centre line of the trench excavation, concrete strip foundation, sub-structure brickwork and dpc, all with a nett length of exactly 75.14 m, which can be used in four separate calculations.

Trench excavation

Volume = centre line × width × depth
= 75.14 × 0.70 × 1 m
= 52.598 m^3

Rounded off: 53 m^3

Strip foundation concrete

Volume = centre line × spread × depth
= 75.14 × 0.70 × 0.150 m
= 7.889 m^3
Rounded off: 8 m^3

Sub-structure brickwork

Area = centre line × height = 75.14 × 1 m = 75.14 m^2
Number of bricks = 75.14 × 120 = 901.68
Rounded off: 902 bricks

Dpc

Linear measurement = centre line + extra for laps

75.14 plus 5% extra for laps = $\dfrac{75.14}{100}$ × 5 = 3.757

Therefore total dpc required = 75.14 + 3.757 = 78.897

Rounded off to 79 m length of 215 mm wide dpc.

STOP

For Example 21 calculate the volume of excavation to form one metre deep by 750 mm wide foundation trenches.

Rounding off – a warning

Rounding up numbers that represent quantities of building materials, when carrying out building calculations, should only be done with the final figure, or last result.

Do not round up two or three times during the working stages of a calculation, otherwise the final quantity could be excessively large, and become what is called an 'accumulated error'.

Circles and triangles

In addition to those shapes of walls and buildings used as examples in this chapter so far, there are two others that the bricklayer will encounter in construction, which require measurement for setting out and calculation; the triangle and the circle, or parts of them.

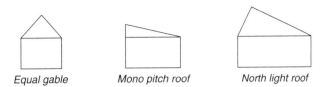

Equal gable   *Mono pitch roof*   *North light roof*

Examples of triangular shape gable end walls

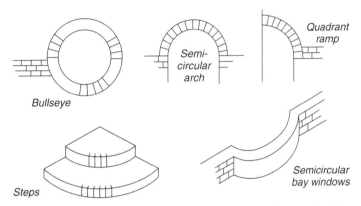

Bullseye   *Semi-circular arch*   *Quadrant ramp*

Steps   *Semicircular bay windows*

Examples of circular shapes or parts of a circle used in construction

Example 30

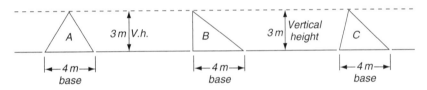

The surface area of any triangle can be calculated as follows, whether shaped like A, B or C.

Surface area = (base × vertical height) ÷ 2
        = (4 × 3) ÷ 2 = 6 m²

The surface area of all three triangles is 6 m², despite their different shapes.

Example 31

Calculate the surface area of the gable end wall for re-pointing.

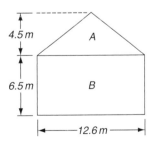

Answer 31

Total surface area = A + B
Area of A = (base × vertical height) ÷ 2
$\qquad$ = (12.6 × 4.5) ÷ 2
$\qquad$ = 56.7 ÷ 2
$\qquad$ = 28.35 m²

Area of B = 12.6 × 6.5
$\qquad$ = 81.9 m²

Total surface area = 28.35 + 81.9 = <u>110.25 m²</u>.

The area of any circular shape can be calculated from the simple formula πR² (pronounced pi R, squared). Some pocket calculators have a button marked π; if not, then use Figure 3.14 (the value for pi never changes). The letter 'R' in the formula stands for radius of the circle.

The perimeter of a circle is called the circumference. Another simple formula is used to calculate the circumference of a circle: πD (pronounced pi, D). π remains 3.14 and 'D' stands for the diameter.

Example 32

Calculate the surface area of the shaded part of the bullseye.

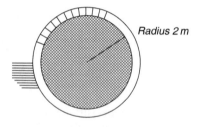

Radius 2 m

Answer 32

Surface area = πR²
$\qquad$ = π × R × R
$\qquad$ = 3.14 × 2 × 2
$\qquad$ = <u>12.56 m²</u>.

Example 33

Calculate the surface area of the circular paved area.

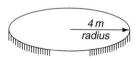

4 m radius

Answer 33

Surface area = πR^2
$= \pi \times R \times R$
$= 3.14 \times 4 \times 4$
$= \underline{50.24 \text{ m}^2}.$

Example 34

Calculate the circumference of the circle shown.

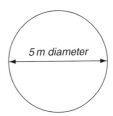

5 m diameter

Answer 34

Circumference = πD
$= \pi \times \text{diameter}$
$= 3.14 \times 5$
$= \underline{15.70 \text{ m}}.$

STOP

Remember that the circumference of a circle is just a plain measurement of length, in linear metres. Surface area is measured in m^2.

Multi-choice questions

This chapter is designed to allow you to check your level of knowledge. It consists of revision questions for Chapters 2–16 in the book. The questions are all multi-choice and have four possible answers. The answers are found at the end of the book.

The main type of multiple choice question will be the type known as the 'four option multiple choice'. This will consist of a question or statement, known as the stem, followed by a choice of four different answers, called the responses. Only one of these responses is the correct answer, the others are incorrect and are known as distracters. You should attempt the questions by choosing either (a), (b), (c) or (d).

Example

The person employed by the local authority to ensure that the Building Regulations are observed is called the:

(a) Clerk of works
(b) Building control officer
(c) Council inspector
(d) Safety officer

a	b	c	d
	✓		

The correct answer is the building control officer, therefore (b) would be the correct response.

2 Materials

Question 1

Which of the following materials could be used to prevent rising damp in a building?

(a) oversite concrete
(b) engineering bricks
(c) foundation concrete
(d) floor screed

Question 2

When building a cavity wall which of the following bricks would be most suitable for the external decorative finish?

(a) engineering bricks
(b) common bricks
(c) facing bricks
(d) fire bricks

Question 3

Bricks that are made by squeezing and extruding a column of clay through a steel die are known as:

(a) wirecut bricks
(b) pressed bricks
(c) hand-made bricks
(d) moulded bricks

Question 4

Which of the following is the correct standard size of a metric brick, without the allowance for joints?

(a) $210 \times 102.5 \times 65$
(b) $215 \times 102.5 \times 65$
(c) $220 \times 102.5 \times 65$
(d) $225 \times 102.5 \times 65$

Question 5

How many bricks are required for the standard brick test to check the dimensional variation?

(a) 24
(b) 25
(c) 26
(d) 28

Question 6

The strength of mortar for brickwork should be

(a) equal to the strength of the brick
(b) weaker than the strength of the brick
(c) stronger than the strength of the brick
(d) any required strength

Question 7

A concrete mix has been prescribed as 1:2:4. Which of the following correctly explains the required mix?

(a) 1 cement   2 lime   4 sand
(b) 1 cement   2 fine aggregate   4 coarse aggregate
(c) 1 cement   2 coarse aggregate   4 fine aggregate
(d) 1 cement   2 sand   4 lime

Question 8

What quality should the water be when used for mixing concrete and mortar?

(a) drinkable
(b) sea water
(c) waste water
(d) any quality

3 Tools

Question 1

Which of the following PPEs is essential when cutting bricks with a club hammer and bolster chisel?

(a) hard hat
(b) ear protector
(c) eye protector
(d) overalls

Question 2

Name the chisel used for cutting out the bed or cross joint in an existing wall.

(a) comb chisel
(b) plugging chisel
(c) bolster chisel
(d) cold chisel

Question 3

Name the item of equipment used for supporting mortar when pointing brickwork.

(a) mortar board
(b) flat hawk
(c) hand board
(d) hand hawk

Question 4

Name the piece of timber marked off with brick courses, used for checking the height of brickwork.

(a) gauge rod
(b) timber rod
(c) pinch rod
(d) black rod

Question 5

A pair of trammel rods are used to set out

(a) curved brickwork
(b) angled brickwork
(c) acute brickwork
(d) obtuse brickwork

Question 6

Name the item of equipment used by the bricklayer for supporting brick-laying lines on a long wall.

(a) corner block
(b) tingle plate
(c) gauge block
(d) storey plate

Question 7

Identify the tool shown below.

(a) brick hammer
(b) skutch hammer
(c) club hammer
(d) claw hammer

Question 8

Identify the item of equipment shown left.

(a) gauge rod
(b) closer gauge
(c) corner block
(d) tingle plate

4 Bonding

Question 1

State the main reason for bonding brickwork.

(a) strengthen the wall
(b) use fewer bricks
(c) make it more decorative
(d) use less mortar

Question 2

Select the most common brick bond for cavity wall construction.

(a) English bond
(b) Flemish bond
(c) stretcher bond
(d) garden wall bond

Question 3

Name the cut brick placed after the corner header in both English and Flemish bonds.

(a) king closer
(b) queen closer
(c) half closer
(d) full closer

Question 4

When the face work is set out with opposite bricks on each course it is known as

(a) reverse bond
(b) half bond
(c) broken bond
(d) walling bond

Question 5

State what you get when you cut a brick in half across its width.

(a) two half bricks
(b) two half bats
(c) two half closers
(d) two half pieces

Question 6

As a general rule of bonding, when you change direction what you should change?

(a) bricks
(b) bond
(c) walls
(d) faces

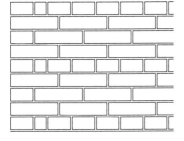

Question 7

Identify the bond shown left.

(a) English garden wall bond
(b) Flemish garden wall bond
(c) stretcher bond
(d) Dutch bond

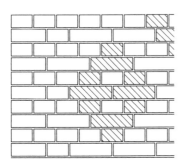

Question 8

Identify the bond shown left.

(a) English garden wall bond
(b) Flemish garden wall bond
(c) stretcher bond
(d) Dutch bond

5 Blockwork

Question 1

Which of the following is used for closing the cavity at a window or door opening?

(a) cut block
(b) reveal block
(c) closer block
(d) special block

Question 2

Name the most common pattern in block walling.

(a) quarter bond
(b) half bond
(c) header bond
(d) block bond

Question 3

Which of the following is the recommended number of blockwork courses to be erected in the same day?

(a) 4
(b) 5
(c) 6
(d) 7

Question 4

Explain the most important reason for using building blocks on the inner leaves of cavity walls.

(a) quickness
(b) cheapness
(c) strength
(d) insulation properties

Question 5

Identify the type of block shown left.

(a) solid block
(b) hollow block
(c) dense block
(d) lightweight block

Question 6

Which of the following is the most common bond used when laying building blocks?

(a) English bond
(b) quarter bond
(c) half bond
(d) Flemish bond

Question 7

Building blocks used below ground level and the full width of the wall are known as

(a) building blocks
(b) breeze blocks
(c) foundation blocks
(d) load bearing blocks

Question 8

The strength of mortar for blockwork should be

(a) equal to the strength of the block
(b) weaker than the strength of the block
(c) stronger than the strength of the block
(d) any required strength

6 Details

Question 1

Name the special brick shown left.

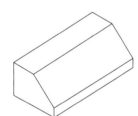

(a) squint brick
(b) plinth brick
(c) radial brick
(d) cant brick

Question 2

Name the bond which consists of alternate courses of headers and stretchers.

(a) Flemish bond
(b) English bond
(c) garden wall bond
(d) stretcher bond

Question 3

Name the bond which consists of alternate headers and stretchers on each course.

(a) Flemish bond
(b) English bond
(c) garden wall bond
(d) stretcher bond

Question 4

A $1^1/_2$ brick wall which is designed to incorporate vertical reinforcement is known as

(a) rat trap bond
(b) quetta bond
(c) Dutch bond
(d) monk bond

Question 5

Name the bond which allows bricks to be laid on edge in a form of Flemish bond.

(a) rat trap bond
(b) quetta bond
(c) Dutch bond
(d) monk bond

Question 6

Name the decorative bond where the bricks are laid with vertical and horizontal joints in line.

(a) single bond
(b) basket weave bond
(c) herringbone bond
(d) stack bond

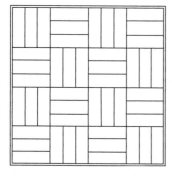

Question 7

Identify the decorative panel shown left.

(a) diagonal basket weave bond
(b) herringbone bond
(c) stack bond
(d) basket weave bond

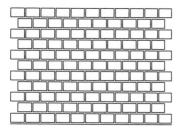

Question 8

Identify the brickwork bond shown left.

(a) stretcher bond
(b) English bond
(c) stack bond
(d) header bond

7 Foundations

Question 1

Name the ground or subsoil on which the building rests.

(a) artificial foundation
(b) natural foundation
(c) concrete foundation
(d) strip foundation

Question 2

What is the distance from ground level to the concrete foundation known as?

(a) foundation depth
(b) atmospheric depth
(c) excavation depth
(d) wall depth

Question 3

State the main purpose of the concrete foundation.

(a) keep the building dry
(b) save brickwork
(c) spread the load of the wall
(d) save mortar

Question 4

The minimum thickness of foundation concrete should be

(a) 100 mm
(b) 150 mm
(c) 200 mm
(d) 250 mm

Question 5

Which of the following is the correct minimum thickness of oversite concrete?

(a) 100 mm
(b) 150 mm
(c) 200 mm
(d) 250 mm

Question 6

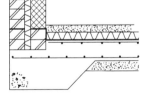

Identify the type of foundation shown left.

(a) strip foundation
(b) raft foundation
(c) short bored piles
(d) wide strip foundation

Question 7

Which of the following is a good type of ground to build on?

(a) peat
(b) silt
(c) hardcore
(d) sandstone

Question 8

Foundations designed to support pillars are called

(a) strip foundations
(b) pad foundations
(c) raft foundations
(d) piled foundations

8 Operations

Question 1

What is the name of the item of equipment used to support bricklaying lines at brick quoins?

(a) tingle plate
(b) corner block
(c) gauge rod
(d) straight edge

Question 2

Name the imaginary line established by the local authority which a building should not encroach.

(a) setting out line
(b) frontage line
(c) building line
(d) side line

Question 3

Which of the following is a right angle?

(a) 1:2:3
(b) 2:3:5
(c) 3:4:5
(d) 4:8:10

Question 4

Name the item of equipment used for checking the accuracy of a right angle.

(a) builder's level
(b) builder's square
(c) builder's edge
(d) builder's rod

Question 5

A gauge rod is used to

(a) check the height of a window
(b) check the height of each course
(c) check the bond is correct
(d) check the perpends are vertical

Question 6

The correct use of a sand course is

(a) for decorative purposes
(b) to allow easy removal of the bricks
(c) to allow the cavity to be ventilated
(d) for re-pointing later

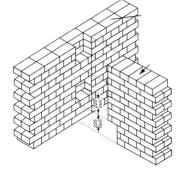

Question 7

Identify the operation shown left.

(a) toothing
(b) racking out
(c) raking back
(d) block bonding

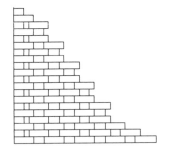

Question 8

Identify the operation shown left.

(a) toothing
(b) racking out
(c) raking back
(d) block bonding

9 Bridging openings

Question 1

What are bricks laid vertical over openings known as?

(a) soldier arch
(b) vertical arch
(c) spanning arch
(d) brick arch

Question 2

Which of the following is the name for bricks in an arch?

(a) key bricks
(b) soffit bricks
(c) voussoirs
(d) soldiers

Question 3

What is the timber support to a semi-circular arch known as?

(a) formwork
(b) springer
(c) centre
(d) turning piece

Question 4

What is the timber support to a segmental arch known as?

(a) formwork
(b) springer
(c) centre
(d) turning piece

Question 5

Which of the following items of equipment can be used for setting out arch curves?

(a) gauge
(b) turning piece
(c) trammel
(d) datum

Question 6

Reinforcement for a concrete lintel should be placed

(a) at the top
(b) just below the top surface
(c) at the bottom
(d) just up from the bottom

Question 7

Which of the following is the name for the slope of brickwork cut to receive an arch?

(a) skewback
(b) springer
(c) closer
(d) voussoir

Question 8

An arch built with wedge-shaped joints is known as a

(a) cut arch
(b) rough arch
(c) axed arch
(d) soldier arch

10 Cavity walling

Question 1

State the main benefit of cavity wall construction.

(a) decoration
(b) insulation
(c) ventilation
(d) condensation

Question 2

Which of the following is the most common bond used for the external leaf of cavity wall construction?

(a) English bond
(b) Flemish bond
(c) stretcher bond
(d) garden wall bond

Question 3

Name the type of wall tie shown left.

(a) double triangle
(b) triangle
(c) butterfly
(d) double V

Question 4

Which is the correct horizontal distance for wall ties in cavity wall construction?

(a) 300 mm
(b) 600 mm
(c) 900 mm
(d) 1200 mm

Question 5

Name the timber used to prevent mortar dropping down the cavity.

(a) cavity strap
(b) cavity trap
(c) cavity rod
(d) cavity batten

Question 6

State the reason for using core holes when constructing cavity walls.

(a) help the cavity to breathe
(b) help to clean the cavity
(c) allow services to enter the cavity
(d) to help insulate the cavity

Question 7

The minimum width of a cavity is

(a) 50 mm
(b) 55 mm
(c) 60 mm
(d) 75 mm

Question 8

The maximum vertical distance of cavity wall ties is

(a) 300 mm
(b) 350 mm
(c) 400 mm
(d) 450 mm

11 Damp proof courses

Question 1

The impervious membrane positioned under all ground floors is known as the

(a) damp proof course
(b) damp proof membrane
(c) damp proof tray
(d) vertical damp course

Question 2

State the name of the bond used to construct sleeper walls.

(a) stretcher bond
(b) honeycomb bond
(c) half bond
(d) quarter bond

Question 3

Which of the following is the minimum height of sleeper walls?

(a) 65 mm
(b) 75 mm
(c) 100 mm
(d) 125 mm

Question 4

Name the term for waterproofing basements.

(a) sealing
(b) bonding
(c) bedding
(d) tanking

Question 5

Moisture entering the building through the external wall is known as

(a) penetrating damp
(b) percolating damp
(c) rising damp
(d) continuous damp

Question 6

Name the material used to weather the tops of walls.

(a) weathering stones
(b) coping stones
(c) paving stones
(d) building stones

Question 7

Why are sleeper walls built with voids between the bricks?

(a) to allow ventilation to pass under the floor
(b) to save in bricks
(c) to save in mortar
(d) to build them quicker

Question 8

State the minimum height of a damp proof course above ground level.

(a) 100 mm
(b) 150 mm
(c) 200 mm
(d) 250 mm

12 Fireplace construction

Question 1

What are the walls either side of a chimney breast known as?

(a) withes
(b) jambs
(c) walls
(d) feathers

Question 2

Name the walls in between multiple flues.

(a) withes
(b) jambs
(c) walls
(d) feathers

Question 3

Which of the following is the correct dimension for projection of a constructional hearth?

(a) 300 mm
(b) 400 mm
(c) 500 mm
(d) 600 mm

Question 4

Name the wall supporting the constructional hearth at ground floor level.

(a) sleeper wall
(b) boundary wall
(c) fender wall
(d) party wall

Question 5

Name the procedure for waterproofing a chimney stack passing through a sloping roof.

(a) tanking
(b) damp proofing
(c) flashing
(d) sealing

Question 6

A flue should have a minimum diameter of not less than

(a) 150 mm
(b) 175 mm
(c) 200 mm
(d) 225 mm

Question 7

Name the material fitted behind a fireback to allow for expansion.

(a) weak concrete
(b) weak mortar
(c) corrugated cardboard
(d) insulation blocks

Question 8

What should be used in the joint between the fireback and the fire surround?

(a) mastic
(b) insulation
(c) asbestos rope
(d) weak mortar

13 External works

Question 1

Name the most effective material for use as a dpc on boundary walls, especially gate pillars.

(a) polythene
(b) engineering bricks
(c) slate
(d) bituminous felt

Question 2

Name the bricks specially manufactured for drives and paths.

(a) kerbs
(b) slabs
(c) pavers
(d) edging bricks

Question 3

Boundary walls can be erected on or near a boundary if they are less than

(a) 600 mm high
(b) 900 mm high
(c) 1000 mm high
(d) 1200 mm high

Question 4

Name the design of the coping stone shown left.

(a) feather edge
(b) saddle back
(c) pointed back
(d) flattened edge

Question 5

Piers built against other walls to give support to the main wall are known as

(a) retaining walls
(b) supporting walls
(c) buttresses
(d) basement walls

Question 6

A brick construction built over channels in a drain or sewer is known as

(a) a catch pit
(b) a cess pit
(c) an inspection chamber
(d) a service chamber

Question 7

A brick bond used in inspection chambers which consists of two leaves each one with contrasting horizontal and vertical joints is know as

(a) English bond
(b) Flemish bond
(c) water bond
(d) manhole bond

Question 8

Identify the pattern shown left used in paving.

(a) diagonal herringbone pattern
(b) running bond
(c) square bond
(d) basket weave bond

14 Special bricks

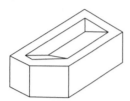

Question 1

Identify the special brick shown left.

(a) plinth brick
(b) cant brick
(c) radial brick
(d) bull nose brick

Question 2

Identify the special brick shown left.

(a) plinth brick
(b) cant brick
(c) radial brick
(d) bull nose brick

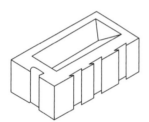

Question 3

Identify the special brick shown left.

(a) frog brick
(b) recessed brick
(c) perforated brick
(d) keyed brick

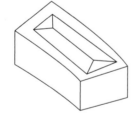

Question 4

Identify the special brick shown left.

(a) squint brick
(b) cant brick
(c) radial brick
(d) plinth brick

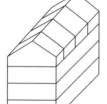

Question 5

Identify the special brick shown left.

(a) squint brick
(b) dogleg brick
(c) radial brick
(d) saddle back brick

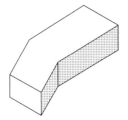

Question 6

Identify the special brick shown left.

(a) squint brick
(b) dogleg brick
(c) radial brick
(d) saddle back brick

Question 7

Plinth bricks are used to

(a) make the wall more decorative
(b) reduce the width of the wall
(c) save on facing bricks
(d) produce a natural curve in the wall

Question 8

Radial bricks are used in the construction of

(a) curved walls
(b) decorative walls
(c) sleeper walls
(d) fender walls

15 Jointing

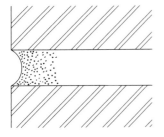

Question 1

Name the joint finish shown left.

(a) flush joint
(b) square recessed joint
(c) half round joint
(d) weather joint

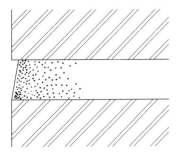

Question 2

Name the joint finish shown left.

(a) flush joint
(b) square recessed joint
(c) half round joint
(d) weather joint

Question 3

Name the tool used for finishing joints shown left.

(a) frenchman
(b) pointing trowel
(c) jointing iron
(d) pointing iron

Question 4

What is the correct depth for raking out mortar joints prior to re-pointing a brick wall?

(a) 6 mm
(b) 12 mm
(c) 18 mm
(d) 14 mm

Question 5

Where is the correct start position when re-pointing a wall?

(a) around doors and window openings
(b) highest point
(c) lowest point
(d) anywhere

Question 6

When the joint finish is completed as the work proceeds it is known as

(a) jointing
(b) pointing
(c) finishing
(d) flushing

Question 7

When the joint finish is applied after the whole area has been completed it is known as

(a) jointing
(b) pointing
(c) finishing
(d) flushing

Question 8

The main purpose of a joint finish is to

(a) make the brickwork more decorative
(b) save on mortar
(c) compact the mortar
(d) make the brickwork stronger

16 Calculations

Question 1

How many bricks are required for a wall 12 courses high and 16 bricks long?

(a) 182
(b) 192
(c) 185
(d) 195

Question 2

A semi-circular centre piece costs £8.45 to make. Calculate the cost of six such centre pieces.

(a) £50.70
(b) £50.07
(c) £57.50
(d) £50.57

Question 3

Add together the following dimensions − 750 mm, 1200 mm, 705 mm and 4645 mm.

(a) 7030
(b) 7300
(c) 3700
(d) 3070

Question 4

A store contains 45 bags of which 24 are cement and the remainder lime. State how many bags of lime are in the store.

(a) 18
(b) 20
(c) 21
(d) 22

Question 5

6000 bricks are required to build a wall of which 80% need to be red and 20% need to be yellow. How many red bricks should be ordered?

(a) 4000
(b) 4600
(c) 4800
(d) 5000

Question 6

Calculate the area of brickwork to be re-pointed if the height of the wall is 3.4 metres and the length of the wall is 32.5 metres.

(a) 110 m²
(b) 110.5 m²
(c) 112 m²
(d) 112.5 m²

Question 7

The distance around an object is known as the

(a) diameter
(b) perimeter
(c) radius
(d) calculation

Question 8

A bricklayer starts work at 8.30 am and after two 15 minute breaks and a 30 minute lunch finishes work at 5.30 pm. How many hours would be paid for a five day week?

(a) 38 hours
(b) 40 hours
(c) 42 hours
(d) 44 hours

Answers to multi-choice questions

2 Materials

1 (b)
2 (c)
3 (a)
4 (b)
5 (a)
6 (a)
7 (b)
8 (a)

3 Tools

1 (c)
2 (b)
3 (d)
4 (a)
5 (a)
6 (b)
7 (b)
8 (b)

4 Bonding

1 (a)
2 (c)
3 (b)
4 (a)
5 (b)
6 (b)
7 (a)
8 (d)

5 Blockwork

1 (b)
2 (b)
3 (c)
4 (d)
5 (b)
6 (c)
7 (c)
8 (a)

6 Details

1 (b)
2 (b)
3 (a)
4 (b)
5 (a)
6 (d)
7 (d)
8 (d)

7 Foundations

1 (b)
2 (b)
3 (c)
4 (b)
5 (a)
6 (b)
7 (d)
8 (b)

8 Operations

1 (b)
2 (c)
3 (c)
4 (b)
5 (b)
6 (b)
7 (d)
8 (c)

9 Bridging openings

1 (a)
2 (c)
3 (c)
4 (d)
5 (c)
6 (d)
7 (a)
8 (b)

10 Cavity walling

1 (b)
2 (c)
3 (c)
4 (c)
5 (d)
6 (b)
7 (a)
8 (d)

11 Damp proof courses

1 (b)
2 (b)
3 (b)
4 (d)
5 (a)
6 (b)
7 (a)
8 (b)

12 Fireplace construction

1 (b)
2 (a)
3 (c)
4 (c)
5 (c)
6 (b)
7 (c)
8 (c)

13 External works

1 (b)
2 (c)
3 (c)
4 (b)
5 (c)
6 (c)
7 (c)
8 (a)

14 Special bricks

1	(b)
2	(d)
3	(d)
4	(c)
5	(d)
6	(b)
7	(b)
8	(a)

15 Jointing

1	(c)
2	(d)
3	(a)
4	(b)
5	(b)
6	(a)
7	(b)
8	(c)

16 Calculations

1	(b)
2	(a)
3	(b)
4	(c)
5	(c)
6	(b)
7	(b)
8	(b)

Index

Brickwork for Apprentices